Production Networks in Asia and Europe

Japanese automobiles dominate the South-East Asian car industry and, although European automobile policies have for a long time been highly discriminatory towards Japanese imports, European makers and suppliers have quickly implemented their production methods. This book examines the various influences of the Japanese automobile industry on industrial development in both South-East Asia and Europe.

In Part I, contributors examine industrial organization and policy issues in Thailand, Malaysia, the Philippines and Indonesia, looking at Japanese investment and the relative policy successes and failures in these host countries. Part II looks at skill-formation systems in the Japanese-dominated automobile industry in South-East Asia and, in Part III, the authors focus on the EU and the very different influence of Japanese investment. In particular, this volume explores:

* how to measure influences on the industrial development process, e.g. human capital formation, rates of technology transfer, borrowing of Japanese systems;
* the increased localization of car production in South-East Asia;
* the adaptability of Japanese models of industrial organization;
* the specific production methods European makers and suppliers have adopted, e.g. Just-in-Time delivery and lean production systems.

These highly original discussions suggest that Japanese assemblers by no means stick to restricted business relations with their traditional suppliers but are more open to cooperation with non-Japanese firms. Both Japanese and European workers in the automobile industry will find this an enlightening read as will those with research interests in Asian business.

Rogier Busser is Head of the Programme for Management Studies, Leiden University and is Coordinator for Academic Research Programmes at the International Institute for Asian Studies.

Yuri Sadoi is Affiliated Fellow at the International Institute for Asian Studies. She previously worked for Human Resources Development at Mitsubishi Motors, Japan.

Sheffield Centre for Japanese Studies/RoutledgeCurzon Series
Series Editor: Glenn D. Hook
Professor of Japanese Studies, University of Sheffield

This series, published by RoutledgeCurzon in association with the Centre for Japanese Studies at the University of Sheffield, both makes available original research on a wide range of subjects dealing with Japan and provides introductory overviews of key topics in Japanese Studies.

The Internationalization of Japan
Edited by Glenn D. Hook and Michael Weiner

Race and Migration in Imperial Japan
Michael Weiner

Japan and the Pacific Free Trade Area
Pekka Korhonen

Greater China and Japan
Prospects for an economic partnership?
Robert Taylor

The Steel Industry in Japan
A comparison with the UK
Hasegawa Harukiyo

Race, Resistance and the Ainu of Japan
Richard Siddle

Japan's Minorities
The illusion of homogeneity
Edited by Michael Weiner

Japanese Business Management
Restructuring for low growth and globalization
Edited by Hasegawa Harukiyo and Glenn D. Hook

Japan and Asia Pacific Integration
Pacific romances 1968–1996
Pekka Korhonen

Japan's Economic Power and Security
Japan and North Korea
Christopher W. Hughes

Japan's Contested Constitution
Documents and analysis
Glenn D. Hook and Gavan McCormack

Japan's International Relations
Politics, economics and security
Glenn D. Hook, Julie Gilson, Christopher Hughes and Hugo Dobson

Japanese Education Reform
Nakasone's legacy
Christopher P. Hood

The Political Economy of Japanese Globalisation
Glenn D. Hook and Hasegawa Harukiyo

Japan and Okinawa
Structure and subjectivity
Edited by Glenn D. Hook and Richard Siddle

Japan and Britain in the Contemporary World
Responses to common issues
Edited by Hugo Dobson and Glenn D. Hook

Japan and United Nations Peacekeeping
Pressures and responses
Hugo Dobson

Japanese Capitalism and Modernity in a Global Era
Re-fabricating lifetime employment relations
Peter C.D. Matanle

Nikkeiren and Japanese Capitalism
John Crump

Production Networks in Asia and Europe
Skill formation and technology transfer in the automobile industry
Edited by Rogier Busser and Yuri Sadoi

Production Networks in Asia and Europe

Skill formation and technology transfer in the automobile industry

**Edited by
Rogier Busser
and Yuri Sadoi**

Routledge
Taylor & Francis Group

LONDON AND NEW YORK

First published 2004 by RoutledgeCurzon
This edition published 2013 by Routledge
2 Park Square, Milton Park, Abingdon, Oxon, OX14 4RN

Simultaneously published in the USA and Canada
by Routledge
711 Third Ave, New York, NY 10017

Routledge is an imprint of the Taylor & Francis Group

Typeset in Baskerville by Wearset Ltd, Boldon, Tyne and Wear

British Library Cataloguing in Publication Data
A catalogue record for this book is available from the British Library

Library of Congress Cataloging in Publication Data
Production networks in Asia and Europe : skill formation and
technology transfer in the automobile industry / edited by Rogier
Busser and Yuri Sadoi.
        p. cm. — (Sheffield Centre for Japanese Studies/
Routledge series)
Includes bibliographical references.
  1. Automobile industry and trade—Japan. 2. Automobile industry
and trade—Europe. 3. Automobile industry and trade—Asia,
Southeastern. 4. Technology transfer—Japan. 5. Technology
transfer—Europe. 6. Technology transfer—Asia, Southeastern.
I. Busser, Rogier Benoãit Pius Marie, 1961– II. Sadoi, Yuri, 1957–
III. Series.
  HD9710.J32P76 2003
  338.4'7629222'094—dc21

                                                    2003005316

ISBN 0-415-31089-X

Contents

Illustrations

Figures

Tables

Box

Contributors

Rogier Busser is working at the Center for Management Studies at Leiden University and at the International Institute for Asian Studies in Leiden.

Ben Dankbaar is Professor of Business Administration at the Nijmegen School of Management of the University of Nijmegen. He has published widely on issues of work organization, organization design and technology and innovation management. He is an expert on the automobile industry.

Jean-Pierre Durand is Professor of Sociology and Director of the Centre Pierre Naville at the University of Paris-Evry. He is the co-author of *After Fordism* (1997) with R. Boyer, and of *Living Labour: Life on the Line at Peugeot-France* with N. Hatzfeld (2003).

Takahiro Fujimoto is Professor at the Division of Economics at the Graduate School of the University of Tokyo and Senior Research Associate at Harvard Business School. He specializes in technology and operations management. His main publications in English include *Product Development Performance: Strategy, Organization, and Management in the World Auto Industry* with Kim B. Clark (1991), and *The Evolution of a Manufacturing System at Toyota* (1999).

Kazuo Koike is former Director of the Kyoto Institute for Economic Research at Kyoto University, and is currently Professor at the Department of Business Studies, Tokai Gakuen University. Among his many publications are *Understanding Industrial Relations in Modern Japan* (1988), *Skill Formation in Japan and Southeast Asia* with T. Inoki (1990) and *Human Resource Development* (1997).

Keisuke Nakamura is Professor at the Institute of Social Science of the University of Tokyo. His most important publication to date is *Nihon no Shokuba to Seisan Shisutemu* (Japan's Production System and the Organization of Work) (1996).

Shinya Orihashi is Lecturer at the Faculty of Economics of Tohoku Gakuin University, as well as a Ph.D. candidate at the Graduate School of Economics, University of Tokyo. His main research area is production systems in the auto industry, especially in foreign subsidiaries of Japanese automotive manufacturers.

Nipon Poapongsakorn is teaching at the Faculty of Economics of Thammasat University, Bangkok, and acts as vice-president of the Sectoral Economics Program at the Thai Development Research Institute in Bangkok. He publishes widely on labour and industrial economics.

Yuri Sadoi is currently affiliated research fellow at the International Institute for Asian Studies at Leiden, the Netherlands. She worked for Human Resources Development at the Mitsubishi Motor Corporation in Japan. Her most important publication so far is *Skill Formation in Malaysian Auto Parts Industry* (2003).

Roland Springer first worked at the Institute for Sociological Research at Göttingen University (SOFI) and later as a senior manager for work organization and labour politics at DaimlerChrysler in Stuttgart. Currently he is Director of the Institute for Innovation and Management in Stuttgart, and gives lectures at Tübingen University.

Joop A. Stam is Professor of Japanese Economic Systems at Erasmus University, Rotterdam and Professor of Management of Technology and Japan at Twente Technical University in Enschede, the Netherlands.

Lepi T. Tarmidi is Professor of Economics at the University of Indonesia and director of the APEC Study Centre of the same university. He publishes widely on Indonesian foreign trade, industrial economics and regional economic integration.

Gwendolyn R. Tecson is Professor at the School of Economics of the University of the Philippines and is currently director for undergraduate studies. Among her main publications are *Catching Up Asia's Tigers*, a two-volume work, co-authored by E.M. Medalla, R.M. Bautista, J.H. Power and associates (1995, 1996). Forthcoming is 'Postwar, Japanese Direct Investments in the Philippines: Trends, Determinants, and Implications for Future Philippine–Japan Relations' in S. Ikehata *et al.* (eds) *Philippines–Japan Relations* (2003).

Thamavit Terdudomtham is Assistant Professor at the Faculty of Economics, Thammasat University Bangkok. His research interest is in Thai industrial economics.

Tham Siew-Yean is Associate Professor and Senior Research Fellow at the Institute of Malaysian and International Studies (IKMAS), Universiti Kebangsaan Malaysia. She publishes on the Malaysian manufacturing sector and on the consequences of globalization and regional integration for the Malaysian economy and ASEAN.

Acknowledgements

This book had its origins in a conference entitled 'New Global Networking in the Automobile Industry: The Effects on Technology Transfer in the Case of Japanese Transplants in Southeast Asia and Europe', organized by the IIAS (International Institute for Asian Studies) in October 2001 in Leiden, the Netherlands. Our appreciation goes to the Japan Foundation, NWO (the Netherlands Organization for Scientific Research), CNWS (the Research School of Asian, African, and Amerindian Studies) and the LUF (Leiden University Fund) for their financial support, which made the conference possible.

The editors would like to thank all of the contributors to this volume for rewriting their original conference papers into chapters for this book and staying on with this production despite their time pressures and busy schedules. With regard to this process, Kamaruding Abdulsomad, Anil Khosla, Yveline Lecler, Thomas Lindblad and Rien T. Segers deserve recognition for their valuable comments in their role as discussants during the conference, offering many suggestions for revisions.

We would also like to express our sincere appreciation to Glenn D. Hook for his support and expert advice that made this publication possible. Our appreciation also goes to the Isaac Alfred Ailion Foundation for its financial support for the publication. Furthermore, we are deeply indebted to Tanja Chute for her highly conscientious work as language editor.

Many of the contributions in this volume are based on extensive field research in the automobile sector, and the publication of this book would not have been possible without the cooperation of those in the automobile industry and their willingness to share their knowledge and experience with the authors of these chapters.

Finally, we would like to thank our families for their continuous encouragement and support.

Rogier Busser
Yuri Sadoi

Abbreviations

ACEA	European Automobile Manufacturers Association
AFTA	ASEAN Free Trade Area
AICO	ASEAN Industrial Cooperation Scheme
AMI	Australian Motor Industries
APEC	Asia-Pacific Economic Cooperation
ASEAN	Association of South East Asian Nations
B2B	business to business
B2C	business to customer
BBC	ASEAN Brand to Brand Complementary Scheme
BOI	Board of Investment (Thailand); Board of Investments (Philippines)
BOT	Bank of Thailand
BUILD	BOI Unit for Industrial Linkage Development
CAD	computer-aided design
CAL	Chrysler Australia
CALS	commerce at light speed
CAM	computer-aided manufacturing
CBU	completely built up
CDP	Car Development Program (Philippines)
CEO	chief executive officer
CEPT	ASEAN Free Trade Agreement Common Effective Preferential Tariff
CIAST	Center for Instructor and Advanced Skill Training (Malaysia)
CKD	completely knocked down
CLEPA	Association of Automotive Suppliers in Europe
CNC	computer numerical control
CNPF	French employers' federation
CVDP	Commercial Vehicle Development Program (Philippines)
DDIT	Double Deduction Incentive (Malaysia)
DOST	Department of Science and Technology (Philippines)
EC	European Commission
EU	European Union
FDI	foreign direct investment
FIMP	First Industrial Master Plan (Malaysia)
FOB	free on board

GATT	General Agreement on Tariffs and Trade
GIAMM	Indonesian Automotive Parts and Components Industries Association
GMHA	GM Holden Australia
GSP	Generalized System of Preferences
HI	heavy industry
HICOM	Heavy Industry Corporation of Malaysia
HRD	human resource development
HRDF	Human Resource Development Fund (Malaysia)
HRM	human resource management
ICT	information and communication technology
IKM	Institut Kemahiran MARA (MARA Vocational Institute)
ISO	International Organization for Standardization
ITAF	Industrial Technical Assistance Fund
ITI	industrial training institution
JAMA	Association of Japanese Automobile Manufacturers
JBIC	Japan Bank for International Cooperation
JICA	Japan International Cooperation Assistance
JIT	Just in Time
KD	knocked down
LCR	local content requirement; local content regulation
LCV	light commercial vehicle
LIUP	Local Industry Upgrading Programme (Singapore)
LMCP	Local Material Content Policy (Malaysia)
MACPMA	Malaysian Automotive Component Parts Manufacturers' Association
MARA	Majlis Amanah Rakyat
MDP	Mandatory Deletion Programme (Malaysia); Motorcycle Development Program (Philippines)
MEDEF	Movement of French Firms
MIDA	Malaysian Industrial Development Authority
MIDI	Metal Workings and Machinery Industry Development Institute
MIT	Massachusetts Institute of Technology
MITI	Ministry of International Trade and Industry (Malaysia; Thailand)
MITP	Malaysia Industrial Training and Productivity
MMAL	Mitsubishi Motors Australia Ltd
MMC	Mitsubishi Motors Corporation
MNE	multinational enterprise
MOSTE	Ministry of Science, Technology and Environment (Thailand)
MSC	MMC Sittipol Co., Ltd
MVDP	Motor Vehicle Development Program (Philippines)
NCP	National Car Project (Malaysia)
NEP	New Economic Policy (Malaysia)
NFEE	net foreign exchange earnings
NIE	newly industrializing economy

NSTDA	National Science and Technology Development Agency (Thailand)
ODA	official development assistance
OECF	Overseas Economic Cooperation Fund of Japan
OEM	original equipment manufacturer
Off-JT	off-the-job training
OJT	on-the-job training
PCMP	Progressive Car Manufacturing Program (Philippines)
PECC	Pacific Economic Cooperation Council
PMMP	Progressive Motorcycle Manufacturing Program (Philippines)
p.p.m.	parts per million
Proton	Perusahaan Otomobil Nasional Sdn. Bhd.
PTMP	Progressive Truck Manufacturing Program (Philippines)
QCC	quality control circle
QCD	quality, cost and delivery
QCDSM	quality, cost delivery, safety and morale
REM	replacement equipment manufacturer
SIMP	Second Industrial Master Plan (Malaysia)
SMEs	small and medium-sized enterprises
SMIDEC	Small and Medium Industries Corporation
SPC	Statistical Process Control
SPM	Malaysian Certificate of Education
SPVM	Malaysian Certificate of Education (Vocational)
TA	technical assistance
TAF	Technology Acquisition Fund
TAI	Thailand Automobile Institute
TAM	Toyota-Astra Motor
TDRI	Thailand Development Research Institute
TFP	total factor productivity
TII	Technology Improvement Institute
TMC	Toyota Motor Corporation
TMCA	Toyota Motor Corporation Australia
TMT	Toyota Motor Thailand
TNSC	transnational supply chain
TPA	(Thai–Japanese) Technology Promotion Association
TPM	total productive maintenance
TPS	Toyota Production System
TQC	Total Quality Control
TQM	Total Quality Management
TRF	Thailand Research Fund
TRIMs	Trade-Related Aspects of Investment Measures
VDP	Vendor Development Programme
WTO	World Trade Organization

1 Introduction

Rogier Busser and Yuri Sadoi

The growth of the Japanese automobile industry in the 1980s was conspicuous for more than the speed of its increase in production volumes. The rapid expansion of overseas production, the integration of these overseas operations into a global production network and the accompanying expansion of integrated global supply networks also attracted attention.

The well-known MIT study *The Machine That Changed the World* (Womack *et al.* 1990) stressed that the Toyota-style 'lean production system' was superior to any other production system in the industry. The overriding sentiment in the automobile sector was that it was hard to fight the Japanese competitors, and a supposedly Japanese model of production became the standard. However, it soon proved to be a mistake to treat all Japanese makers as equal. In the late 1990s, Nissan, Mitsubishi and Isuzu were in financial trouble and large capital investments from European and American makers were necessary to rescue these companies. Toyota and Honda, on the other hand, were still going strong and experienced far fewer difficulties.

Despite these difficulties, it is clear that the globalization of Japanese automobile makers has influenced considerable changes in the automobile industries in those countries that host Japanese investments in this sector. Moreover, even in Europe, where Japanese automobile investment has been limited in terms of capital flows, Japanese models of production influence European makers. A large body of literature has treated many different aspects of these influences on Asia, Europe and the United States.[1]

In this volume, the focus is on production networks and industrial organization. It is precisely through these two mechanisms that Japanese enterprises have extended important influences upon Asian and European automobile industries and indirectly upon governmental automobile industrial policies. The case studies in this book make clear that when Japanese industrial organization models are transferred abroad, they influence modes of technology transfer and systems for local skill formation. At the same time, however, Japanese industrial organization models show a strong ability to adapt themselves to local conditions.

This choice to study the automobile industry in terms of technology transfer from Japanese enterprises to foreign enterprises and the Japanese influence upon skill-formation systems can be argued from many perspectives. In the first place, the automobile industry has a wide variety of technologies in use, ranging from simple assembly and plastic moulding to state-of-the-art robotic welding technologies. In the second place, and rather unsurprisingly, governments in both developed and developing countries perceive the automobile sector as an important means by which to upgrade their industrial structure. To this end, many countries have adopted automobile-specific industrial policies. Last, but not least, the sheer size of the automobile sector makes it difficult to ignore. Since its beginnings, the automobile industry has created new models in industrial organization. The American domination of the industry, exemplified by the well-known term 'Fordism', continued into the 1970s. In that decade, however, fuel-efficient Japanese cars started to conquer overseas markets while still using mass production systems. In the 1980s, Japan witnessed a gradual shift away from mass production and towards lean production systems. The rising influence, or even dominance, of Japanese automobile producers is illustrated by the term 'Toyotism' that emerged in the late 1980s (Dohse *et al.* 1985; Kadoto 1985; Nomura 1993; Shimizu 1999). The subsequent internationalization of the Japanese automobile industry that took off in the 1980s, and still continues, further justifies the approach of this study in terms of technology transfer from Japanese to foreign enterprises, and Japanese influences upon skill-formation systems.

This internationalization of the Japanese automobile industry was triggered by two external but interrelated developments. First, in the early and mid-1980s, high levels of automobile exports from Japan resulted in more severe import regulations in most major Japanese overseas markets. Second, the revaluation of the yen after the Plaza Accord 1985, which in itself was partly a consequence of high export levels of automotive products, forced Japanese automobile manufacturers to cut their production costs. One way of doing so was to accelerate production volume overseas. The major Japanese producers quickly succeeded in this process: the number of vehicles produced outside Japan increased from 1 million in 1986 to 6 million in 1998, while exports from Japan decreased from 7 million in 1986 to 5 million in 1996, as shown in Figure 1.1. The second way to cut production costs was sought through improving the production system itself. Improvements were pursued by widely implementing lean production systems, which first took effect in Japan proper and was soon after attempted at the overseas production sites of Japanese manufacturers.

The lean production system was the Japanese automobile industry's most important comparative advantage. However, lean production systems cannot easily be transferred abroad. Even if the production facilities and processes (hardware) are transferred, to operate effectively the organizational structures also have to be transferred. Many studies were conducted on the Japanese approach to the organization of production (Boyer *et al.* 1998; Freyssenet *et al.* 1998). These studies agree that the major strengths of the lean production system are smooth production flows combined with product variety, consistently high conformance to standards and continuous improvement of both the product and the production processes (Takahashi 1998).

First, the lean production system requires a specifically designed model of human resource management. This model aims at the efforts of managers

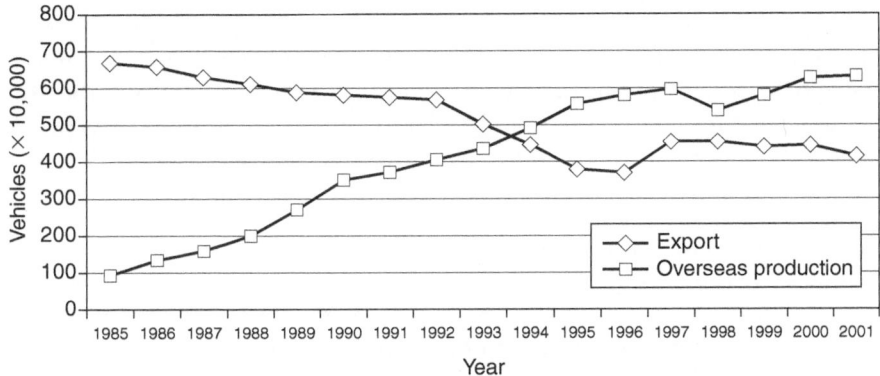

Figure 1.1 Production volume of Japanese automobiles, 1985–2001

Source: Japan Automobile Manufacturers Association (2002)

and engineers to innovate, and to maximize workers' capability on the shop floor. The lean production system was developed over many years under the Japanese human management system and in a favourable social and cultural environment. The skills and competence required for a specific job are most effectively learned at the workplace and include empirical knowledge, operational skills and a high theoretical competence. These skills are developed through enterprise-based training, especially on-the-job training (OJT), which is the dominant system for skill formation in Japan (Koike and Inoki 1990; Koike 1997).

Second, the lean production system depends on a specific type of industrial organization, i.e. the way in which supply networks are structured, organized and maintained by major automobile makers. Toyota, Nissan and, to a lesser extent, Mitsubishi had established and developed vertical supply networks that could deliver parts and components at very competitive prices while maintaining high quality standards. The organization of the supply networks was designed in such a way that Just-in-Time (JIT) delivery systems could be implemented among all cooperating firms. Moreover, the supply network enabled a smooth implementation of the lean production system in the supplying firms, as well. In order to achieve a JIT system that functioned well, enterprises strongly emphasized human resource development (HRD) through instituting systems such as quality control circles and creating learning organizations. Again, these systems were implemented not only in final assembly firms like Toyota, Nissan, or Mitsubishi, but in all enterprises involved in the supply networks.

As the lean production system is dependent on a specific type of industrial organization, the transfer of production technology abroad by Toyota, Nissan and Mitsubishi required the formation of local supply networks similar to those in Japan. In the late 1980s, local suppliers both in the United States and in South-East Asia were not available to Japanese final manufacturers. In the

United States, this was because the Big Three produced high ratios of parts and components in-house, while in South-East Asia technological levels of the local suppliers were of a poor standard. Consequently, Japanese final assemblers requested their suppliers to invest in production facilities in the United States and in several South-East Asian countries. In this way, a number of important features of Japanese industrial organization were transplanted abroad. This study does not focus on the United States, but rather on the consequences of this transplanting of principles of Japanese industrial organization for companies in South-East Asia and Europe.

In this study, the choice of Japan as an exporter of a model of industrial organization is not confined to the success of Toyota, which often serves as the example of Japanese industrial organization. The Japanese auto industry still has considerable competitive strengths, despite financial difficulties and alliances with foreign partners. Moreover, the issues of technology transfer, skill transfer and national policies in industrial development are attracting more widespread interest and the auto industry is still a major industry.

South-East Asia and Europe have been chosen to offer a contrast between one region where Japanese automobile producers are very dominant, South-East Asia, and another region where Japanese market share and production have been rather limited, Europe. In fact, the production volumes of Japanese makers in Europe are far below those of Japanese makers in the United States (Figure 1.2). Moreover, while numerous studies have been published on the Japanese automobile industry in North America, few have appeared on the influence of the Japanese automobile industry on Europe and European companies.

It goes without saying that a true comparison between the influence of Japan experienced by the automobile industries in South-East Asia and Europe respectively cannot be made because the automobile industries in these regions differ too much. However, by studying Japanese industrial organization and network formation in South-East Asia and Europe, we can see how Japanese companies adapt themselves to local circumstances and how mutual influences gradually develop into new models of industrial organization.

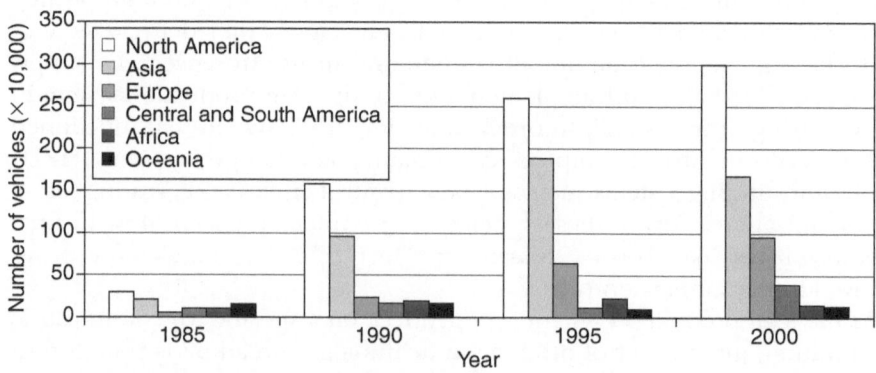

Figure 1.2 Japanese automobile production by areas

Source: Japan Automobile Manufacturers Association (2002)

ASEAN automobile policies and the role of the Japanese automobile industry

ASEAN, established in 1967, has not yet succeeded in developing a common policy for the automobile industry. Thus when we discuss ASEAN automobile policies, we are in fact referring to policies of individual member states. Even before ASEAN was established, Thailand, Indonesia and the Philippines had already attempted to develop automobile industries. Within the greater industrial policy framework, the automobile industry was part of a broader import substitution developmental strategy. Under this strategy, most South-East Asian countries introduced high tariff barriers and made use of several non-tariff barriers to support the growth of a local automobile industry. Thailand, for example, introduced high import tariffs on imported cars in 1962 and tax benefits for completely knocked down (CKD) production. By raising import tariffs in 1967 and increasing CKD production in 1969, Malaysia aimed to reach a 20 per cent local content level.

Under these import substitution policies, CKD car production increased in all South-East Asian countries during the 1970s. However, governments soon realized that the benefits of this industry for their domestic economies were very limited because assembly companies were either Japanese, joint ventures, or local companies producing under licence, and in all three cases were heavily dependent on technology supplied from Japan and were generating little value added. Moreover, the automobile sector in South-East Asia remained highly dependent on imports of parts and components. This situation resulted in big deficits on bilateral trade balances with Japan because the need for more inputs for the production of the sector increased in line with the growing demand for automobiles.

To raise a strong class of local suppliers and to push for technology transfer, most ASEAN countries developed so-called local content policies. The basic idea was to force car manufacturers to purchase locally either certain parts and components or a set percentage of the total value of a car. If final assemblers met these local content regulations (LCRs), the cars would be considered locally made and be taxed favourably. These policies were introduced in the early 1980s at a time when ASEAN economic policies were shifting from import substitution to more outward-looking, or export-oriented, policies. However, only since 1986, when the impact of the Plaza Accord became apparent, have Japanese investors responded positively to this new policy. As Japanese automobile manufacturers increased their production volumes overseas (Figure 1.3), Japanese parts suppliers started to establish overseas production sites near their Japanese customers at overseas locations. In cases in which the suppliers were already active before 1986 in South-East Asia, investments and increased production volumes were stepped up. Thus, in 1986–1987, not only did investments by Japanese final assemblers grow quickly, but investments from parts suppliers also increased. The role of local enterprises as suppliers to Japanese final assemblers or first-tier suppliers remained, on average, limited.

Although ASEAN did not develop an integrated policy for the automotive sector, some special schemes and regulations were designed to support its

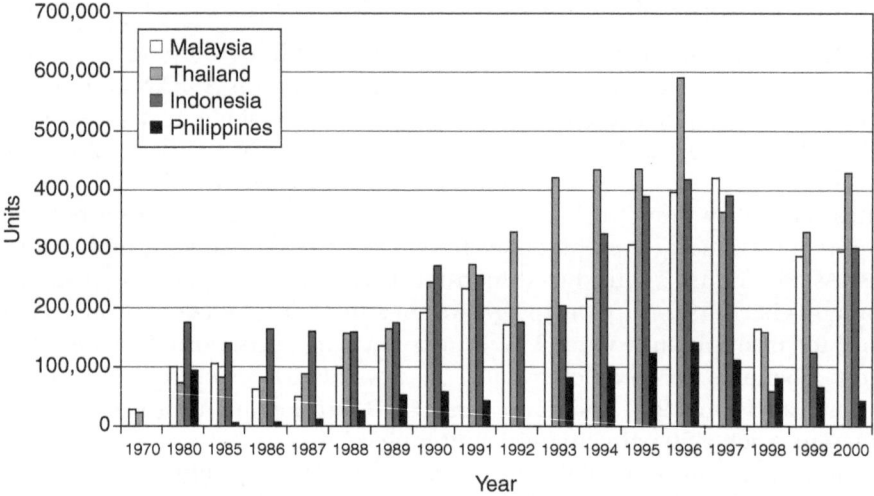

Figure 1.3 Automobile production volume in ASEAN countries

Source: Nikkan Jidosha Shinbunsha (1994–2002)

further development. For example, the Brand-to-Brand Complementary Scheme (BBC Scheme) was introduced in October 1988. This scheme made it possible for final assemblers within ASEAN to exchange parts within the network of operations of one single final assembler and receive a 50 per cent discount on import duties, but only when parts were traded under the brand name of the final assembler. For example, Toyota could exchange parts manufactured by Toyota Thailand with parts manufactured by Toyota Philippines. There were also cases when parts produced by local suppliers were exchanged after a final assembler had attached its brand name to it. The plan stipulated that a part produced in one of the ASEAN countries is treated as locally produced with respect to all ASEAN countries. This was important for meeting local content standards. Although the BBC Scheme never became a success, it was a breakthrough in the sense that it was a first attempt in ASEAN towards an integrative process for the automobile industry.

Largely because of the criticism that non-ASEAN companies, in particular the Japanese, benefited most from the BBC Scheme, it was replaced in 1996 by the ASEAN Industrial Cooperation Scheme (AICO Scheme). This is a mechanism that further promotes joint manufacturing activities among companies based in ASEAN. Under the AICO Scheme, companies can benefit from the lower Common Effective Preferential Tariff (CEPT). The CEPT is, in fact, a step towards a fully fledged ASEAN free trade area. It reduces tariffs to between 0 and 5 per cent in 2003 and stipulates that the original six ASEAN members should eventually eliminate all duties on products by 2010. A serious obstacle to a smooth implementation of these plans, and thus for further integration of the automobile industry in ASEAN, is Malaysia's policy to put its own automobile industry on an exclusion list.

The ASEAN economies developed very well in the years after 1986. Purchasing power for many South-East Asians increased and the automobile market was booming. Japanese brands had captured between 80 and 95 per cent of most South-East Asian markets. Now that these markets were growing, scale economies became apparent and accordingly Japanese final assemblers and partmakers stepped up investments in the second half of the 1980s and the early 1990s.

The financial crisis of 1997 that started in Thailand and soon spread to most other ASEAN countries had serious repercussions for the automobile industry. Sales fell, and consequently production was put back or, in some cases, even temporarily halted. Japanese dominance of the automobile industry in South-East Asia remained firm, although American and European makers gained some market share in the period following the Asian crisis. The crisis deepened Japanese involvement in the ASEAN automobile sector because Japanese investors came up with additional investments in existing joint ventures to save these endangered enterprises.

European Union policies and the role of the Japanese automobile industry

The European automobile industry was originally developed by German, British, French and Italian enterprises. With the ongoing process of European integration, however, national automobile markets have become increasingly interdependent over time. Moreover, the European automobile industry has become more internationalized, owing to ever-increasing flows of trade in goods and services and, more importantly, through the dynamics of international investment, product innovation and technology transfer.

The sudden rise of Japan as an important car-manufacturing nation in the early 1970s and the expansion of Japanese automotive exports to Europe affected European automobile enterprises. The two oil crises of the 1970s brought the Japanese car industry to the fore of international automobile markets, as Japanese makers tended to produce the most fuel-efficient cars. In the two decades that followed the 1970s, Japanese cars gained ever-better reputations, as they were perceived by European customers to be both reliable and of good quality for the price.

During the 1970s and 1980s, Japanese automobile makers increased both their production volume and their exports. In 1970, 126,275 vehicles were exported from Japan to European countries, comprising 11.6 per cent of total exports from Japan. By 1975, the exports from Japan had increased fourfold and reached 528,486 vehicles. They continued to increase and reached 1,226,954 vehicles in 1980 and 1,363,694 vehicles in 1990. In the two decades between 1970 and 1990, the number of cars exported from Japan to Europe increased nearly tenfold. European countries became an important export destination for Japanese automobile manufacturers. In 1970, only 12 per cent of exports from Japan were for the European market, but by 1990 the proportion had risen to 30 per cent (Table 1.1).

The 1990s was a decade that brought import regulation under the European monitoring policy and the rise of Japanese automobile production in

Table 1.1 Exports of Japanese automobiles

Year	North America	Europe	C. and S. America	Africa	Middle East	Asia	Oceania	Others	Total
1965	40,404	16,458	14,117	21,595	8,400	56,879	36,176	139	194,168
1970	495,608	126,275	79,698	111,244	26,635	149,787	87,316	233	1,076,796
1975	1,003,954	528,486	143,509	217,294	241,511	290,134	251,426	1,298	2,677,612
1980	2,592,577	1,226,965	382,231	322,329	542,955	581,116	316,865	1,934	5,966,972
1985	3,384,563	1,363,694	290,417	137,729	401,598	710,573	426,075	15,809	6,730,458
1990	2,521,821	1,750,497	216,375	129,293	284,194	569,143	344,236	15,994	5,831,553
1995	1,301,218	918,831	329,064	137,718	206,446	616,027	274,828	6,676	3,790,808
2000	1,836,941	1,136,083	298,801	110,218	295,176	410,599	357,739	9,337	4,454,894
2001	1,795,816	895,415	293,556	98,524	381,965	351,227	341,808	7,778	4,166,089

Sources: JAMA (2002); Nikkan Jidosha Shinbunsha (1994)

Europe. In view of the importance of the automobile sector, many European governments supported their domestic industries and took refuge in trade measures. There was a rapid escalation in unofficial import restrictions and Europe forced voluntary export restraints upon Japanese automobile makers (Mulhearn *et al.* 2001).

As the number of imports from Japan intensified, the market share of Japanese automobiles in Europe increased from around 5 per cent in the early 1980s to over 10 per cent in 1991. Germany had a 36.9 per cent share of the market in 1991, the largest in Europe, followed by France (23.4 per cent), Italy (13.6 per cent), Japan (10.1 per cent) and the United Kingdom (7.8 per cent) (Nikkan Jidosha Shinbunsha 1994).

Japanese car sales to Europe levelled off at the end of the 1980s, as the major producing European countries limited Japanese imports. Italy restricted them to a minuscule 2,550 cars per year, although up to 40,000 may have been imported from elsewhere in Europe. France set a limit of 3 per cent of its market for Japanese cars. Under pressure, the Japanese agreed to limit sales of their cars in the British market to about 11 per cent and in the West German market to about 15 per cent (Laux 1992). The tariffs and import regulations for Japanese cars in Europe in 1994 are shown in Table 1.2.

The European Commission (EC) had earlier decided to remove all quotas on automobiles by the mid-1990s. However, this decision was reversed in 1991 when European carmakers made demands for the continuation of barriers against Japanese cars until the end of 1999. Then, in 1992, the EC reached an agreement with the Japanese Automobile Manufacturers Association (JAMA) on the gradual lifting of import restrictions on Japanese cars by 2000. According to this agreement, Japan was allowed to increase its exports from 1994 onward. Table 1.3 shows how monitory (meaning 'closely watched') numbers of Japanese exports to EU countries in the period 1993–1999 increased gradually from 980,000 to 1,245,000 units.

The automobile industry has traditionally been subject to trade intervention, but the instruments, as well as the scope, of the actions are gradually

Table 1.2 Tariff and import regulation for Japanese cars

	Tariff for PCs (%)	*Tariff for CVs (%)*	*Import regulation*
EU	10	11–22	No
Germany	10	11–22	No
UK	10	11–22	No
France	10	11–22	Regulation for Japanese cars
Netherlands	10	11–22	No
Belgium	10	11–22	No
Italy	10	11	Regulation for Japanese cars
Spain	10 only for Japanese cars		Regulation for Japanese cars
Portugal	10	11–22	Regulation for Japanese cars
USA	2.5	25	VRA for Japanese cars

Source: Nikkan Jidosha Shinbunsha (1994)

Note
PCs: passenger cars; CVs: commercial vehicles; VRA: Voluntary Restraint Agreement

Table 1.3 Monitory number of Japanese exports to EU countries, 1993–1999

	1993	1994	1995	1996	1997	1998	1999
UK	202,800	183,100	182,700	180,000	188,000	190,000	190,000
France	69,000	74,900	88,700	99,700	94,200	111,100	114,700
Italy	38,800	47,000	55,300	65,500	102,200	107,000	129,500
Spain	29,300	32,400	40,400	44,500	56,400	70,000	92,700
Portugal	39,000	39,500	32,800	34,000	34,500	43,000	50,000
Others	601,100	607,100	671,100	653,300	638,700	668,900	668,100
Total	980,000	984,000	1,071,000	1,077,000	1,114,000	1,190,000	1,245,000

Source: Fourin (2000)

shifting and are closely related to the relocation of production facilities and the establishment of transplants. Since the 1980s, Japanese automobile makers have established production plants in Europe. Table 1.4 shows Japanese automobile production plants in Europe. First, Nissan started production in Spain in 1983 and in the United Kingdom in 1986. Honda started production in the United Kingdom in 1989. Following that, Toyota also started its operations in the United Kingdom in 1992. Besides these examples, Japanese automobile makers also currently run another seven plants in continental Europe.

Owing to the increase of local production in Europe by Japanese producers, the imports of Japanese passenger cars from Japan are below the limits of the monitory imports of the Japanese car. This observation especially applies to France, Italy and Spain (Fourin 2000).

The number of Japanese automobile production facilities has climbed since the early 1980s. Production volume has also increased. However, the market share of the combined Japanese producers has remained at around 11 per cent of the total cars registered in Europe. Table 1.5 shows the number of imported cars from Japan and the locally produced cars in Europe from 1995 to 1998. During these four years, roughly 26 per cent of the Japanese passenger cars and less than 3 per cent of Japanese commercial vehicles sold in Europe were locally produced. Thus while Japanese brands have a market share of 11.3 per cent in Europe, only 23.7 per cent of this amount was actually produced in Europe. These figures imply that only

Table 1.4 Japanese automobile production in Europe

Country	Maker	Year production started	Equity	Number of employees	Production volume per year
UK	Nissan	1986	Nissan 99%	4,900	300,000
	Honda	1989	Honda 100%	3,000	150,000 cars 200,000 engines
	Toyota	1992	Toyota 100%	3,450	200,000 cars 200,000 engines
Netherlands	Mitsubishi	1995	Mitsubishi 50%	5,500	200,000
France	Toyota	2001	Toyota 100%	2,000	150,000
Portugal	Toyota	1968	Toyota 27%	1,955	14,000
	Mitsubishi	1972	Mitsubishi 99%	362	12,000
Spain	Nissan	1983	Nissan 99.7%	3,900	150,000 cars 100,000 engines 82,000 T/M
Poland	Toyota	2002	Toyota 100%	300	250,000 T/M
Turkey	Toyota	1994	Toyota 40%	856	32,000
	Honda	1998	Honda 100%	300	30,000

Sources: JAMA (2002); Fourin (2000)

Note
T/M: transmissions

Table 1.5 Sales of Japanese cars in Europe

	1995		1996		1997		1998		Total (1995–1998)		
	PC	CV	PC	CV	PC	CV	PC	CV	PC	CV	Total
Import	1,007,838	174,540	982,534	170,539	1,107,155	202,536	1,244,657	238,549	4,342,184	786,164	5,128,348
Local production	280,323	5,089	412,342	6,038	437,369	5,846	444,419	5,653	1,574,453	22,626	1,597,079
Total	1,288,161	179,629	1,394,876	176,577	1,544,524	208,382	1,689,076	244,202	5,916,637	808,790	6,725,427
Rate of local production	21.8%	2.8%	29.6%	3.4%	28.3%	2.8%	26.3%	2.3%	26.6%	2.8%	23.7%
Share of registration	10.7%	11.6%	10.9%	10.8%	11.5%	11.8%	11.8%	12.2%	11.2%	11.6%	11.3%

Source: Fourin (2000)

Note
PCs: passenger cars; CVs: commercial vehicles

2.6 per cent of all cars registered in Europe were produced by Japanese automakers in Europe. In the case of passenger cars, the proportion was nearly 3 per cent. Compared to South-East Asian markets, but also to the American market, these figures are very low. Thus although the Japanese automobile industry has certainly influenced that of Europe, its production in Europe is still quantitatively small.

Structure of the book

This volume consists of fifteen chapters. After this Introduction, the book is divided into three parts. The first deals with industrial organization and policy issues in ASEAN, the second focuses on the problem of the transferability of 'Japanese' skill formation systems and the third part concerns the question of Japanese influence upon European models of industrial organization and skill formation systems.

Part I deals, as already mentioned, with industrial organization and policy issues in ASEAN. These topics are linked to Japanese investment in the sector and the relative successes and failures of their host-country policies are analysed. For South-East Asian countries, the success of automotive policies might be measured in terms of technology transfer. Most authors rely on original field research in their arguments that technology transfer takes place and is stimulated by (local) industrial policies, but the authors also point out that dependence on foreign (Japanese) technology remains high.

Stam's introductory chapter to Part I of the volume offers a number of theoretical concepts on industrial organization and technology transfer. These concepts are linked to the realities of East and South-East Asian industrial development and offer an outline of the major characteristics of the East Asian and South-East Asian industrial structures, but also provide the theoretical underpinnings for the following four chapters on the automobile industry in Thailand, Malaysia, the Philippines and Indonesia respectively. These country studies are clearly written from a developmental perspective.

Terdudomtham analyses the evolution of Thai government policies for the development of the automobile industry in Thailand. He focuses on the role of government policy and agencies in technology transfers from (foreign) automobile assemblers to local auto partmakers. His chapter offers original evidence that government policies have had positive effects on technology transfer. However, Terdudomtham also concludes that the Thai automobile sector is still insufficiently developed to be competitive in the globalized auto parts industry.

Tham assesses the impact of Malaysia's automotive policies on the development of this sector, including its technological development. She analyses the policy options available for Malaysia in view of increasing pressures to liberalize and globalize the automobile industry. Her findings indicate that, while national policies have resulted in (a) the first national car (Proton) capturing a considerable portion of the local market, and (b) the development of local automotive components and parts-manufacturing enterprises, there is still a pronounced dependence on foreign technology, especially with regard to engine design.

Tecson examines the evolution in Philippine government policies for the automobile industry from the 1970s to the present. She focuses on the direct and indirect implications of such industrial policies for technology transfer by multinational companies operating in the Philippines and for human capital formation. She points out that there was no systematic government policy that targeted technology transfer. A number of case studies show the difficulties that confronted local assemblers and suppliers.

Tarmidi discusses Indonesian industrial policy for the automobile sector with a focus on technology transfer. Since the early 1980s, the Indonesian government has tried various policy measures to increase the share of locally made components, with the ultimate goal of establishing an Indonesian automobile industry. Under this policy, the achievement of higher local content was rewarded with lower import tariff rates. However, this policy was, on the whole, not very successful. Indonesian companies remain highly dependent on foreign technologies, and local production is often based on licensing agreements.

The first part of the book concludes with a chapter by Fujimoto and Orihashi. This contribution investigates how differences in resource endowments and dynamic capabilities of firms from the same home country (Japan) affect the strategic and competitive behaviour of their operations in the same foreign country. The analysis concentrates on two Japanese auto assemblers, Toyota and Mitsubishi Motors, which both have production facilities in Australia and Thailand. Because the automobile sector has received a severe shock in recent years in both these countries, they provide interesting case studies. These show that differences in firm size in the home country affect their overseas policies, choices and strategies.

The second part of the book focuses on the problem of the transferability of 'Japanese' skill formation systems. The authors present original research with emphases on case studies conducted in factories in Japan, Indonesia, Thailand and Malaysia. The results of these case studies represent important new findings and contribute to a better understanding of skill formation, work organization, human resource management and the interdependence of these factors.

Koike's introduction to Part II concerns the question as to whether the Japanese HRD system can be transferred abroad. Koike starts his analysis by disclosing the vital character of workers' skills in Japanese industry. Examining the transferability of the Japanese system requires defining what the 'Japanese system' means. If, for example, the Japanese system heavily depends on so-called dense networks between personnel without any high level of technical skills, then one can only conclude that chances for its transferability are low. If, on the contrary, the Japanese system emphasizes high levels of technical skills, then chances for its transferability are high. According to a series of field studies in contemporary Japanese industries, it is intellectual skill that most promotes competitiveness. Based on the research results on twenty-eight workshops in Toyota and its related firms, Koike's chapter discloses the essential character of intellectual skills, how such skills can be built up and their prospects under the development of information technology and robotization.

Poapongsakorn addresses some important issues of labour training in the Thai automotive industry and offers an analysis of the role Japanese automotive producers have in implementing the Japanese employment system and their influence on training patterns, skill formation and skill upgrading in the Thai automotive industry. The extensive fieldwork reveals that education and training are complementary and that training, in itself, does not directly enhance workers' earnings. As better-educated workers are more likely to be trained by their employers, Poapongsakorn suggests that the most efficient government intervention is investment in higher education.

In Chapter 10, Sadoi examines the auto parts industry in Malaysia in order to understand the difficulties a developing country faces when promoting skill formation. The Malaysian government has promoted automobile production and its supporting industries by giving tax benefits and tariff protection through a national car project. Despite a history going back nearly two decades, the technical capabilities of workers in the Malaysian auto industry are still low. The increase in local content is taken as an indicator of the progress of the Malaysian auto industry. Although local content has increased, this is largely due to extensive imports of capital goods that facilitate difficult production process for parts production. Case studies of technicians in forging and casting are used to illustrate difficult areas of skill acquisition.

Nakamura approaches his analysis of the international transfer of technology from the perspective of work organization and applies it to the transfer of the Toyota Production System (TPS) to Toyota-Astra Motor (TAM), a Japanese–Indonesian joint venture that has operated in Indonesia for about thirty years. The case study presented shows that the sophisticated Japanese production management technology has been successfully transplanted into TAM and, moreover, that Toyota's flexible production system is being executed in TAM. Nakamura also maps out the work organization that supports this flexible production system and explains how human resource management (HRM) in TAM improves workers' skills.

The third and final part of this volume focuses on the EU. For European countries, the influence of Japanese investment in the automobile industry has been of a different sort. European automobile policies have long been highly discriminatory towards Japanese automobile imports. Although the market share of Japanese cars is limited, Japanese automobile production has affected European production systems in several ways. European makers and suppliers quickly implemented innovative Japanese production methods such as the 'lean production system'. In addition, work organization and skill formation were modified to facilitate smooth implementation of the lean production system.

Dankbaar's introduction to Part III provides an overview of skill-formation systems in the European automobile industry. A series of case studies on training in European countries shows that it has become a key element in a far-reaching process of restructuring that is currently under way in the industry. He discusses the organizational changes and their influences on the training needs. Moreover, he examines training and knowledge transfer between carmaker and their suppliers and discusses European approaches to skill formation.

Busser and Sadoi's chapter examines the formation and structure of supply systems at Japanese car manufacturers in Europe. The case studies indicate that the relationship between Japanese car manufacturers in Europe and their Japanese and European parts suppliers in Europe is very different from the relationship that Japanese car manufacturers have with their suppliers in Japan. Busser and Sadoi also show that Japanese suppliers did not invest in continental Europe primarily to supply their major Japanese customers, but started production in Europe to gain market share, to learn from the demands of European customers and to gain competitiveness.

Springer provides a theoretical analysis of the influences of the lean production system on work organization in the German automobile industry. This contribution acknowledges the importance of flexibility to the lean production system but, on the other hand, Springer points out that flexibility has reached its limit and is obstructing any further development of the German automobile industry. Therefore, he makes a plea to return to flexible standardization and indicates how new job designs can support this.

Some important Japanese influences on work organization in the French automobile factories are discussed by Durand. The French car industry experienced profound difficulties in the first half of the 1980s, whereafter the industry developed methods of adaptation to international competition and enhanced its position in Europe, as well as in Asia and some of the areas covered by Mercosur. Profitability has been enhanced. Durand's chapter illustrates how two French auto producers were able to achieve these results.

Note

1 For studies on the Japanese influence on Asia, see Shin (1999), Tokunaga (1992) and Ogawa and Makido (1990).

References and further reading

Boyer, R., Charron, E., Jürgens, U. and Tolliday, S. (eds) (1998) *Between Imitation and Innovation: The Transfer and Hybridization of Productive Models in the International Automobile Industry*, Oxford: Oxford University Press.
Cusumano, M.A. and Takeishi, A. (1991) 'Supplier Relations and Management: A Survey of Japanese, Japanese-Transplant, and U.S. Auto Plants', *Strategic Management Journal*, 12: 563–588.
Dohse, K., Jurgens, U. and Malsch, T. (1985) 'From "Fordism" to "Toyotism"? The Social Organization of the Labour Process in the Japanese Automobile Industry', *Politics and Society*, 14: 2.
Fourin (1999) *1999 Oshu Jidosha Buhin Sangyo* (1999 Auto Parts Industry in Europe), Nagoya: Fourin.

—— (2000) *2000 Oshu Jidosha Sangyo* (2000 Automobile Industry in Europe), Nagoya: Fourin.

Freyssenet, M., Mair, A., Shimizu, K. and Volpato, G. (eds) (1998) *One Best Way? Trajectories and Industrial Models of the World's Automobile Producers*, Oxford: Oxford University Press.

Fujimoto, T., Nishiguchi, T. and Ito, H. (eds) (1998) *Sapuraiya Shisutemu* (Supplier System), Tokyo: Yuhikaku.

Ishida, M., Fujimura, H., Hisamoto, N. and Matusmura, F. (1997) *Nihon no Riin Seisan Hoshiki* (The Japanese Lean Production System), Tokyo: Chuo Keizaisha.

Japan Automobile Manufacturers Association (JAMA) (1999) *1999 The Motor Industry of Japan*, Tokyo: JAMA.

—— (2002) 'Motor Vehicle Statistics' (online). Available at http://www.jama.or.jp (accessed 15 December 2002).

Kadoto, Y. (1985) *Toyota Shisutemu* (Toyota System), Tokyo: Kodansha.

Koike, K. (1997) *Nihon Kigyo no Jinzai Ikusei* (Human Resource Development in Japanese Firms), Tokyo: Chuko Shinsho.

Koike, K. and Inoki, T. (eds) (1990) *Skill Formation in Japan and Southeast Asia*, Tokyo: University of Tokyo Press.

Laux, J.M. (1992) *The European Automobile Industry*, New York: Twayne Publishers.

Mulhearn, C., Vane, H. and Eden, J. (2001) *Economics for Business*, Basingstoke, UK: Palgrave.

Nikkan Jidosha Shinbunsha (1994–2002) *Jidosha Sangyo Handobukku* (The Auto Industry Handbook), Tokyo: Nikkan Jidosha Shinbunsha.

Nomura, M. (1993) *Toyotizumu* (Toyotism), Tokyo: Mineruba.

Ogawa, E. and Makido, T. (1990) *Ajia no Nikkei Kigyo to Gijyutsu Iten* (Japanese Enterprises in Asia and Technology Transfer), Nagoya: Nagoya University Research Center.

Shimizu, K. (1999) *Le Toyotisme*, Paris: La Découverte.

Shin, D. (1999) *Trust in Lean Production Systems: Lean Job Design and Workers' Trust in Management at Korean Automobile Plant*, Paris: Insead Euro-Asia Centre Research Series.

Takahashi, Y., Murata, M. and Rahman, K. (eds) (1998) *Management Strategies of Multinational Corporations in Asian Markets*, Tokyo: Chuo University Press.

Tokunaga, S. (1992) 'Japan's FDI Promoting Systems and Intra-Asia Networks: New Investment and Trade Systems Created by the Borderless Economy', in Tokunaga, S. (ed.) *Japan's Foreign Investment and Asian Interdependence*, Tokyo: Tokyo University Press.

Womack, J.P., Jones, D.T. and Roos, D. (1990) *The Machine That Changed the World*, New York: Rawson Associates.

Part I

Industrial organization and industry policy in ASEAN

2 Industrial organization, culture and technology transfer

Joop A. Stam

In 1990, Womack, Jones and Roos published their book *The Machine That Changed the World*. This report of the MIT study on the international competitiveness of the US, European and Japanese car industries was the starting point of a broad campaign for lean production, continuous improvement and all the other techniques that generated a worldwide interest in Japanese business management.

Now, more than a decade later, Japanese industry, and particularly the car industry, is only a shadow of its former self. Among all the proud carmakers, only Toyota and Honda are more or less independent, i.e. they can make their own decisions. All the others, such as Mazda, Mitsubishi, Nissan, Suzuki and Subaru, have substantial foreign participation from US or EU car manufacturers. Japanese commentators euphemistically tend to refer to this foreign participation as 'business alliances to reduce cost', but, in effect, most Japanese carmakers have lost their say. Nissan is run from Paris, Mazda is supervised by Ford and Mitsubishi has to comply with the rules of Daimler-Chrysler. The deplorable state of the Japanese economy, the rock-bottom stock prices, and the current international crisis are failing to generate the optimism that Japan and the Japanese car industry so much need.

However, there is still a lot to discuss about Japan and Japanese industry, in particular in the domain of technology, technology management, technology transfer and technological cooperation. In this introductory chapter to the first part of the book, I would like to focus on the industrial organization of Japanese mass manufacturing, the encompassing business culture and their combined impact on technology transfer.

To start with the last of the above, the issue of technology transfer became very much alive when Japanese companies changed their business strategy and started to manufacture goods abroad instead of producing in Japan and exporting them to overseas markets. In the early stages of this process, there were some quite serious misgivings about the added value of Japanese factories. 'Screwdriver factories' was the first assessment, in that they created employment but not high-quality employment. Most Japanese factories that had been set up abroad used Japanese systems and operated under the supervision of Japanese staff and technicians. Local personnel were hired for the routine assembly jobs.

Since the early 1980s, this picture has changed in Japanese companies' favour. Now, electronic, car, machine and other types of industries are welcomed everywhere because of their positive effects on investment in the local economy, industry and society. In 1991, Womack, Jones and Roos published the second edition of their book *The Machine That Changed the World*, but with the added subtitle *How Japan's Secret Weapon in the Global Car Wars Will Revolutionize Western Industry*. For that matter, the revolution would also come to industry in East and South-East Asia.

In 1995, the Pacific Economic Cooperation Council (PECC) concluded that foreign direct investment (FDI) 'may provide an even stronger stimulus to economic development of the region than increased trade', because of the expected spillover effects of technology-related investment in local production facilities. More precisely, technology transfer generated by FDI would improve the technology base of the host country and consequently its economic viability. In this respect, Berri and Ozawa (1997) refer to a process of 'comparative advantage recycling' – that is, as the industrialization of East and South-East Asia proceeds, a structural upgrading of technological sophistication takes place, from low-tech and labour-intensive manufacturing processes to hi-tech and capital-intensive ones. Of course, this upgrading through technology transfer concerns Japanese as well as Western FDI in the Pacific Asia region.

What do we mean by *technology transfer*? Many definitions are circulating, but here I use a definition derived from Agmon and Glinov (1991). Technology transfer concerns the acquisition, development, utilization and maintenance of knowledge, including technical knowledge (hardware, software and humanware), and takes place between governments, institutes, enterprises and individuals. Thus technology transfer can be the result of aid agreements between governments as, for instance, through official development assistance (ODA) or exchange arrangements between institutions such as universities. For the purpose of this discussion, I primarily concentrate on technology transfer at enterprise level.

As regards the car industry, I refer to technology transfer between enterprises, namely, the Japanese car manufacturers and the domestic firms of the host countries in Europe and Pacific Asia. I am particularly concerned with the effects of the interaction between the Japanese car manufacturers and the domestic firms, partners and either suppliers or subcontractors.

Within Japan, substantial technology transfer takes place between the core firm and the suppliers and subcontractors in the vertical production chain, the so-called *sangyo keiretsu*, the pyramid of loyal small and medium-sized companies supplying the parts and components to the final assembler. This Japanese network structure for mass production provides the possibility for the core firm to reap the benefits of specialization and flexible manufacturing within the companies belonging to the network, all of which, in turn, results in the reduction of production costs.

One of the primary conditions for realizing these benefits has been close cooperation between the core firm and its main suppliers. Although the relationship cannot be labelled egalitarian, the interdependence between the core company and its suppliers has had a beneficial effect on inter-company technology transfer. In order to secure the required quality and quantity of parts, components and sub-assemblies, Japanese carmakers have invested in the training and education of personnel at the supplying companies, often also providing financial assistance for their equipment. Knowledge about both production techniques and production management has thus been shared. Within this structure, long-term relations are a prerequisite for the joint design of new products and the joint development of product markets. The sharing of all necessary information available within the network enables the cluster of companies to enter new markets and reduce risks. Collaboration in research and development (R&D) often follows in due course.

This cooperative structure has yielded the high-quality, cost-competitive products that have become the hallmark of Japanese industry and created its competitive edge in the world market. It seems obvious that Japanese enterprises, manufacturing either in Pacific Asia or Europe, are trying to transfer this model of cooperation to their transplants.

To what extent can this pattern of cooperation, which worked so well in Japan, be repeated in Europe and Pacific Asia and will it remain successful under the conditions of global competition and global networking? It is from that perspective that I would like to discuss Japan's integrated manufacturing systems, with particular attention to the following topics:

- the changing economic and industrial environment of the car industry;
- the changing nature of the Japanese industrial networks; and
- the technological structure of Japanese enterprises in Pacific Asia and technology transfer.

The changing economic and industrial environment

In the early 1990s, Japan's car industry had a strong position in world markets. After the Plaza Accord of 1985 and the resulting rise of the yen, large-scale investment took place in production facilities in both the Pacific Asian region and Europe. In Europe's case, the announcement of the unification of the internal market in 1992 triggered extra investment in factories within the border of the Union. In Pacific Asia, there emerged an international division of labour. Those parts of the Japanese industrial production process with a low added value or that were considered dangerous and dirty had been transferred to the respective host countries in East and South-East Asia where they could benefit from welcome packages, preferential tax treatment, or cheap local labour. Step by step we have witnessed the substitution of the export of finished goods from Japan by local production in East and South-East Asia destined for local markets and eventually also for overseas markets. However, those parts of the Japanese business system with a relatively high value or strategic importance, such as R&D, design, and marketing, have remained in Japan and are directly supervised by corporate headquarters.

As Japanese manufacturing companies moved their assembly activities to South-East Asia, in particular Malaysia, Thailand, Indonesia and the

Philippines, they were confronted with problems related to the supply of parts and components. Local industry in the region was not always able to provide parts and components in the quantities and qualities required. Subsequently, the Japanese assembling companies invited their own well-known suppliers and subcontractors from Japan to set up shop in Asia and thus guaranteed themselves a steady flow of parts to the assembly lines.

This policy has had two clear side effects. On the one hand, the transfer of assembly processes to Asia, including the main suppliers and subcontractors, has led to a hollowing out of the Japanese manufacturing industry in Japan, *kudoka*. This reduction of industrial employment has been felt sharply in areas where these suppliers and subcontractors have been concentrated, usually the countryside. So far, the economic slowdown of the 1990s has delayed the replacement of the disappearing industries with new employment.

On the other hand, setting up a manufacturing industry in Pacific Asia together with loyal and faithful subcontractors meant, in essence, the replication of the Japanese business structure in Pacific Asia. More importantly, those Japanese suppliers and subcontractors who followed their patron could enjoy preferential treatment. Consequently, they have become a serious impediment, if not a threat, to the development of local suppliers and subcontractors in South-East Asia, contrary to the original idea that local industry would benefit from these Japanese investments.

Within a period of ten years, all major manufacturing affiliates of the Japanese car and electronics industry have established themselves in Pacific Asia and are organized in clusters and webs around the assembly plants of Toyota, Mitsubishi, Sony, Matsushita and all the others. Of course, this increased investment resulted in employment for the local people, income for the local shopkeepers, tax for the government and a general upgrading of the industrial environment, but the effect on the industrial and technological development of the host country is highly disputed. Hatch and Yamamura (1996) conclude that Asian countries are not lifting off under their own steam but are pulled forward, in part by Japanese capital and technology or, more explicitly, by their membership in a Japanese production alliance. They call this 'embraced development'.

In spite of this structural impediment that hinders local industry from benefiting fully from Japanese FDI, the continuous growth and development of the Asian region up to 1997–1998 generated many opportunities for local Asian companies too. Since the early 1990s, trade and investment barriers have been substantially lowered as a result of the Asia-Pacific Economic Cooperation (APEC) and World Trade Organization (WTO) policies. Consequently, local Pacific Asian companies could use their prime competitive advantage and low labour costs, and start to export their products, e.g. to Japan. So while many Japanese manufacturing affiliates have moved to the Pacific Asian region, competitive Korean, Taiwanese, Singaporese and recently also Chinese producers of parts and components are invading the Japanese home market. Thus Japanese industry is facing what it likes to call 'mega-competition' from its Asian neighbours and it has to adjust quickly to this new fact of life.

To summarize, the international division of labour in Asia, actively promoted by Japanese industry, the globalization of business and the deregulation of the product markets within the APEC and WTO policy frameworks have significantly changed the landscape of Japanese manufacturing industry and resulted in a major reorientation of structure and organization.

The changing nature of the network

The inter-firm networks, the *sangyo keiretsu*, of the manufacturing industries in Japan are currently the targets most haunted by the prospect of change. At the top of the pyramid, we come across the overstaffed assembly companies with excessive and almost uncontrollable personnel costs. Even R&D of new technology and products seems to be stalling, but the most dramatic changes are taking place in the *keiretsu* network of suppliers and subcontractors. Nissan, the number two carmaker of Japan, is just such an example. Carlos Ghosn, the new French CEO of Nissan, decided to clean up the system and sell many of the shares of the various subcontractors that Nissan owned. This divestment not only meant an alleviation of Nissan's financial burden (most of the shares were not performing well, and Nissan too was performing badly), but also entailed a break with the intra-network traditions and practices.

In his book *Information and Organization: A New Perspective on the Theory of the Firm* (1997), Mark Casson associates business networks and networking with social groups or local communities. The essence of a network is that different members trust each other. Trust is naturally stronger within small groups than large ones because of the frequency of face-to-face meetings, but trust can also be found in national or even international networks. The importance of a network lies not so much in the smooth flow of goods or services within the network, as in the flow of information that is used to *coordinate* the material flow. Thus the essence of a network is that it accommodates a distinctive high-trust method of coordination, and in that way it reduces the transaction costs.

This is exactly what we have observed in the past in the Japanese production network, the *sangyo keiretsu*. The long-term relations between the core manufacturer and subcontractors based upon mutual trust and economic benefit have greatly facilitated the coordination of complex production processes in the assembly industries. Breaking down this trust-related system as a means to rationalize economic processes, as we see, for instance, at Nissan, cannot continue without consequences. Responsible people within Nissan realize this. Therefore, we should not expect a wholesale sell-out of all former 'trust-related relations', but probably we will see more selective behaviour towards the large group of suppliers and subcontractors. Only those with high added value to the core company will continue to enjoy preferential relations, while an increasing part of the pyramid will be subjected to market forces where price competitiveness will be the name of the game. Furthermore, competition will no longer come only from within the industry itself, but also from other Asian competitors. It will be very interesting to see to what extent this trend in competition will also affect the Japanese production networks in Pacific Asia. Will we see a shift to more 'market' in the subcontracting relations, away from the pattern of exclusiveness and favouritism, and will this mean a better chance for the local producers?

At this juncture, a few additional remarks seem justified about the emerging new relationship between the core company and its preferred coterie of subcontractors. In the past, effective management and coordination mechanisms have been a strong competitive advantage for Japanese companies; however, according to Porter *et al.* (2000), strategic choice and new technology have become the prime factors for survival in the global environment. When new technology has been jointly developed by the core company and its suppliers, and thus has been embedded in the network, one imagines that the relationship between them will most likely remain strongly influenced by this mutual dependency. This mutual dependency, or 'institutional lock-in', might have negative effects for both carmakers and their partmakers. The exclusive relation of Japanese carmakers with their specialist partmakers could prevent them from freely shopping around outside the network for the much-needed advanced technology. Conversely, partmakers may be prevented from exploiting their expertise and going global. That was also the experience of Nippon Denso, the high-profile electronic equipment maker for the Toyota network. When Denso sold the jointly developed engine management technology without Toyota's consent, it was punished by Toyota by exclusion from the development of the new hybrid car, the Prius. Another effect of participating in an exclusive network is *adaptive engineering*. Within the electronic industry in South-East Asia, we can observe that the exclusive relationship with a Japanese assembly company, using its patents, licences, equipment and production technology, quickly leads to a strong technological dependency and reduces the possibility for the local company to grow and expand beyond the network.

Technology structure and technology transfer

Among the many assumptions about FDI and technology transfer is that they are economically beneficial to the host country but also a remedy against industrial backwardness. To what extent can that idea be maintained? At this point, I would like to make some assessments of technology transfer based on Mitsuhiro Seki's model of technological structure.

Seki (1994) depicts the technological structure of a country or economy as a triangle and distinguishes three layers. The bottom layer of the triangle consists of *fundamental or base technologies*, such as, for instance, casting, forging, pressing, machining and all those techniques and skills that enable us to produce tools, utensils and simple machines in an industrial way. In the second layer, he positions *intermediate technologies*, or the skills and techniques required to assemble more complicated products in mass production processes from cars to computers. The top of the pyramid consists of *advanced technologies*, or special technologies, the fruits of R&D and scientific discoveries. Space technology, nuclear technology and biotech, among others, belong to this category. On the basis of this classification, we can generate a technology structure profile for each country or economy and make an assessment of its strengths and weaknesses, needs and chances for successful technological cooperation.

When we apply Seki's model to the Pacific Asian region, we can say the following: Japan has a broad base of fundamental technologies. Both its historical legacy in craftsmanship and 150 years of industrial experience have provided this fundament. In addition, during the past fifty years, Japan has developed a set of intermediate technologies that have become the international standard in mass manufacturing. Finally, Japan has reached a well-developed level of advanced technologies, as for instance in electronics, nuclear technology and space science.

South Korea and Taiwan, both former colonies of Japan, have well-developed fundamental technologies, although not so broadly based and deeply embedded, since they started to industrialize fully only in recent times. During the past three decades, they have proved able to master the skills of mass production, but their level of achievement in advanced technologies is still limited and requires substantial investment.[1] As for the ASEAN countries, until the end of their colonial period they had a predominantly agricultural economy and a weak industrial base. After they gained independence, the initial phase of industrialization was characterized by gradual improvement in their fundamental technologies. But since the 1980s, ASEAN countries such as Thailand, Malaysia, Indonesia and the Philippines have started a process of accelerated industrial development through FDI and the utilization of their labour cost advantage in mass manufacturing processes. FDI has been the key to their achieving their current level of technological capabilities. Without having well-developed fundamental technologies of their own, they manage to master intermediate technologies through assistance programmes and imports from Japan and the West. Advanced technologies seem still somewhat remote, although investments in science and technology are increasing (Malaysia, like South Korea, made an offer to take over Fokker).

In China, we come across well-developed fundamental technologies, the result of a history of more than three thousand years of advanced craftsmanship. However, since Chinese society became collectivized under communist rule in the 1950s, intermediate technologies for mass consumption goods have not had priority. Only from the mid-1980s onwards have intermediate technologies and the production of mass consumption goods been given full attention again. In the process, China has made ample use of Japanese and Western FDI to fill the technological gaps. Advanced technologies, on the contrary, have always been well positioned in China, the result of the country's geopolitical strategy to match its own military technologies with those of other world powers.

On the basis of these profiles, we may state the following: Korea and Taiwan enjoy a well-balanced development of their technological capabilities and most likely will make further progress through FDI, alliances and their own investment in science and technology. Availability of human and financial resources will probably be the main factors influencing the speed of development.

The ASEAN countries have based their growth strategy on intermediate technologies imported through FDI and they try to internalize them as quickly as possible. In order to succeed and become technologically independent, they need to improve their industrial base of fundamental technologies or accept large-scale imports of parts and components from abroad. Japan's policy to transfer its integrated production systems to the region could be beneficial to the ASEAN countries, but only when these systems have remained in place for a long time or when there are real technological spillovers from the networks of Japanese suppliers and subcontractors.

China probably holds the best cards for future technological and industrial achievement. At the moment, Japanese and Western investment in mass production facilities is quickly filling the gap in intermediate technologies. To judge from the scale and speed at which it is proceeding, this process will soon be completed and China will become an equal player in the technology game, and it will probably be able to take advantage of its giant labour reserve for a long time to come.

Japan has opted for an international division of labour and, since the end of the 1980s, has been actively transferring parts of its manufacturing industry to the Pacific Asian region. As a consequence, its home base in fundamental technologies is getting weaker and the prolonged recession of the 1990s has not improved the situation. At the same time, it tries to stay ahead of the surrounding countries in technological competence by investing in R&D but, as Japan's own past experience makes clear, it is much harder to be the leading party than to catch up with others.

Conclusion

In this introductory chapter to the first part of the book, I have broadly sketched the state of affairs in Japanese manufacturing industry, and its current problems. Continuous pressure for change from inside and outside have forced the Japanese car manufacturers to adjust their industrial organization and reconsider their culture of cooperation. Since Japanese business cannot dominate the game any more, they have to venture into alliances, thereby hoping to learn how to survive the global competition.

Note

1 When the Dutch aircraft maker Fokker was in financial trouble and on the brink of bankruptcy, the Korean government and Korean industry were interested in taking over the company and all its technological assets in an attempt to speed up the process of catching up in special technologies. Eventually the deal failed to materialize.

References

Agmon, T. and Glinov, M.A. (1991) *Technology Transfer in International Business*, Oxford: Oxford University Press.

Berri, D.J. and Ozawa, T. (1997) 'Pax Americana and Asian Exports: Revealed Trends of Comparative Advantage Recycling', *International Trade Journal*, 11 (1) (Spring): 39–67.

Casson, M. (1997) *Information and Organization: A New Perspective on the Theory of the Firm*, Oxford: Oxford University Press.

Hatch, W. and Yamamura, K. (1996) *Asia in Japan's Embrace: Building a Regional Production Alliance*, Cambridge: Cambridge University Press.

Porter, M., Takeuchi, H. and Sakakibara, M. (2000) *Can Japan Compete?*, Cambridge, MA: Perseus Publishing.

Seki, M. (1994) *Beyond the Full-Set Industrial Structure: Japanese Industry in the New Age of East Asia*, Tokyo: LTCB International Library Foundation.

Womack, J.P., Jones, D.T. and Roos, D. (1990) *The Machine That Changed the World*, New York: Rawson Associates.

3 Thai policies for the automobile sector

Focus on technology transfer

Thamavit Terdudomtham

The automobile industry in Thailand suffered drastic changes as a result of the 1997 economic crisis, trade liberalization and changes in the global automobile industry, including the investment of US auto assemblers in Thailand.[1]

Following the economic crisis and sharp reduction of domestic sales in 1997–1998, auto assemblers in Thailand reduced production and became more export oriented (see Tables 1.3, 3.1 and 3.2). With the abandonment of local content requirements (LCRs) in 2000, auto assemblers employed a global part-sourcing strategy more extensively. In addition, US assemblers brought the market-based bidding system for outsourcing to Thailand.[2]

The auto part industry in Thailand has also changed drastically since the latter half of the 1990s. Several global auto partmakers followed the US auto assemblers and invested in Thailand. Because of the economic crisis, the elimination of foreign ownership restrictions by the Board of Investment

Table 3.1 Domestic vehicle sales in Thailand

Year	Passenger cars	Commercial cars	Total
1985	22,097	63,125	85,222
1986	22,481	55,973	78,454
1987	27,116	74,508	101,624
1988	38,768	107,712	146,480
1989	47,705	160,538	208,243
1990	65,864	238,198	304,062
1991	66,779	201,781	268,560
1992	121,441	241,546	362,987
1993	174,169	282,299	456,468
1994	155,670	330,008	485,678
1995	163,371	408,209	571,580
1996	172,730	416,396	589,126
1997	132,060	231,096	363,156
1998	46,300	97,765	144,065
1999	66,858	151,472	218,330
2000	83,106	179,083	262,189
2001	104,502	168,639	297,052

Sources: Thai Automotive Industry Association, Federation of Thai Industries and Thailand Automotive Institute

Table 3.2 Export of automobiles from Thailand

Year	Passenger cars	Commercial cars	Total	Percentage change
1987	488	40	528	
1988	14,121	314	14,435	2,634
1989	9,032	631	9,663	−33
1990	5,826	442	6,268	−35
1991	15,688	144	15,832	153
1992	2,384	1,586	3,970	−75
1993	3,781	8,532	12,313	210
1994	14,294	7,086	21,380	74
1995	1,623	7,182	8,805	−59
1996	926	15,493	16,419	86
1997	564	41,654	42,218	157
1998	6,773	59,999	66,772	61
1999	n.a.	n.a.	125,702	85
2000	n.a.	n.a.	152,835	22
2001	n.a.	n.a.	175,299	15

Sources: The Customs Department of Thailand, and Thailand Automotive Institute

(BOI) and the abandonment of LCRs, some local first-tier suppliers were taken over by Japanese or Western firms. Some of them were downgraded to become second-tier suppliers or lower, some shifted to other businesses. In the midst of their struggle, one of the major problems encountered by these local auto partmakers was that their technological capability, especially with regard to product design, was insufficient.

This chapter aims to examine the Thai government's policy on the development of its domestic automobile industry. To this end, the first section offers a comprehensive overview of the restructuring of the automobile sector in Thailand and the second focuses on the role of government policy with regard to technology transfer[3] in the automobile industry.

Restructuring and modernization of the automobile industry

Major structural changes of the automobile industry in Thailand, which began in the late 1980s, have been accompanied by an increase in both the number of assemblers and parts suppliers, and changes in the ownership structure and modernization process. This section will first describe these structural changes and then analyse the factors affecting the restructuring process.

The development of the automobile industry began with investment incentives in 1962, which attracted twelve foreign auto firms and some Japanese parts firms to Thailand. Rapid development under an import substitution strategy, however, accelerated only after a series of policy revisions that were aimed at market rationalization, and the local market experienced rapid growth in the early 1970s. Consequently, the number of assemblers at this time fell to eleven firms, when GM pulled out of Thailand. This number remained the same until the BOI began a new round of promotions in 1994, after which the number of promoted assemblers increased to sixteen firms. The increase was in response to the rapid economic growth in the period 1987–1993 and the liberalization policy in the early 1990s. The last firm to be promoted was BMW, which received the BOI certificate in 1998. Therefore, not counting two truck assemblers, there are presently fifteen assembly firms, seven of which still had the promotional certificates in 1999 (BOI 1999a). Two of the new entrants, GM and Ford, are from the American Big Three, and established their plants in the late 1990s. Chrysler, the other Big Three company, entered the assembly business in Thailand after it merged with Daimler-Benz, since the latter had already had an assembly plant in Thailand for more than two decades (Table 3.3).

Although the auto assemblers were foreign investors, those Western and Japanese assemblers had entered the Thai market via links with large Thai conglomerates, some of which had close relations with Thai bankers. A study by Doner (1991: 47) showed the local participation to have been fairly strong and even to have expanded between 1972 and 1985.

Local capital also played an important role in the parts industry. Doner's study found that ten of the twelve firms established during the 1960s and at least thirty parts firms established between 1962 and 1975 were completely Thai owned (1991: 191). They initially emerged to supply the replacement parts market, and thus are known as 'replacement equipment manufacturers' (REMs). Most of the Thai parts producers acquired their know-how from technical links with foreign producers. Several local parts firms led the movement of parts firms for greater stimulation of local industry in the 1970s and 1980s (ibid.: 47).

As a result of the rapid growth of the auto market and the moderately successful policy supporting local industry in the 1970s, 'the Thai parts firms increased from several dozen in 1970 to over 200 by the mid-1980s; 150 of these were original equipment manufacturers (OEMs)' (ibid.: 47). The industrial boom in the late 1980s together with the liberalization policy in the early 1990s led to a jump in auto sales, attracting a large number of new entrants into the parts industry. Various studies estimated that there were about 700–800 parts firms in 1994–1996 (see Table 3.4). If the small-scale firms in metal working, moulds and dies and electronics parts are included, there might have been as many as 2,000 firms in 1996 (Poapongsakorn 1997) and 3,000 firms in 1998 (*Krungthep Thurakit* newspaper, 28 August 1998). These numbers seem reasonable given the fact that there are 1,211 auto parts companies listed in the 2000 Auto Parts Directory in Thailand. Out of the 700 parts producers in the mid-1990s, 490 firms were REMs and 210 companies were OEMs (Brooker Group 1997). Most of the OEMs were medium-scale Japanese firms or joint ventures. After the economic crisis in 1997, a large number of Thai-owned parts suppliers, particularly SMEs, went bankrupt. Although there have been some new foreign entrants since 1998, the current number of parts producers is probably smaller than 700 firms. Unfortunately, there are no official data.

Table 3.3 Production capacity of the automobile industry in Thailand, 2001

	Assembler	Passenger cars	Pick-ups	Trucks and buses	Total
1	Toyota Motor Thailand (1964)	100,000	100,000	—	200,000
2	MMC Sittiphol (1966)	42,000	118,000	14,400	174,400
3	Isuzu Motor Thailand (1966)	—	110,000	20,000	130,000
4	Siam Nissan Automobile (1963)	—	78,000	3,900	81,900
5	Honda Cars Manufacturing Thailand (1993)	50,000	—	—	50,000
6	Siam Motors and Nissan (1963)	31,200	—	—	31,200
7	Bangchan General Assembly (1972)	20,000	—	—	20,000
8	Thonburi Automotive Assembly (1956)	13,500		1,400	14,900
9	YMC Assembly (1975)	12,000		—	12,000
10	Thai Rung Union Car (1973)	9,600		—	9,600
11	Hino Motor Thailand (1964)			9,600	9,600
12	Thai Swedish Assembly (1976)	6,000		—	6,000
13	Siam VMC Automobile		6,000	—	6,000
14	Motor and Leasing (Thailand)			200	200
15	Auto Alliance Thailand (1995)		201,000	—	201,000
16	General Motors Thailand (2000)	40,000	—	—	40,000
17	BMW Manufacturing (Thailand) (2000)	10,000	—	—	10,000
	Total	334,300	613,000	49,500	996,800

Sources: Ministry of Industry (2001), BOI (2001) and Terdudomtham (1997)

Note

The years in which assembly started are given in parentheses

Table 3.4 Number of parts suppliers in Thailand

Year	Total	OEM	REM	Thai (new firms)	Japanese		Including SMEs in other, related industries
					All firms	New firms	
1960s	n.a.	—	—	10	—	12	—
1970	Several dozen	—	—	—	—	14[a]	—
1977	112	—	—	—	—	7[b]	—
Mid-1980s	200	150	—	—	—	11[c]	—
1994–1996	700–800	210	490	—	84	65[d]	2,000
1998	1,211	—	—	—	—	—	3,000
2001	Less than 700	—	—	—	—	—	—

Sources:
1 1960 to mid-1980s from Doner (1991).
2 1994–1996 from Poapongsakorn (1997) and the Brooker Group (1997).
3 1998 from *Krungthep Thurakit* newspaper (28 August 1998) and *2000 Auto Parts Directory in Thailand*.
4 Data for new firms from Poapongsakorn (1997).

Notes
a Number of new firms established during 1971–1975.
b Number of new firms established during 1976–1980.
c Number of new firms established during 1981–1985.
d Number of new firms established during 1986–1994.
OEM, original equipment manufacturer; REM, replacement equipment manufacturer; SMEs, small and medium-sized enterprises

Changes in ownership structure and strategies of MNEs

Despite strong local participation in the automobile industry, Japanese investors have played a key role from the beginning and have increasingly dominated the industry since the late 1980s. In the 1970–1980 period, six of the eleven assemblers were Japanese and they had a combined market share of more than 90 per cent (see Figure 3.1).

There were two major changes in the late 1990s. Thanks to liberalization and the ASEAN Free Trade Area (AFTA), American and Korean assemblers decided to enter the Thai market. Although the market share of the American Big Three is still relatively small, it has increased significantly since 1996 (Figure 3.1). Moreover, their presence has had a significant impact on both the structure of the parts industry and competition in the car market. Japanese assemblers have responded with a new concept of a cheap Asian car model, e.g. the City by Honda and the Soluna by Toyota.

The second change was caused by the economic crisis in 1997, which led to a sharp decline in car sales. The financial difficulties forced the foreign partners to inject more capital into the assembly firms, which resulted in marginalization of the Thai shareholders. For example, Toyota Thailand had to increase its capital by 5 billion baht, which was ten times its original capital. As a result, Siam Cement, which used to be the major shareholder, decided to hold on to only a small share in Toyota Thailand.[4] The crisis, therefore, has substantially reduced Thai capital in the assembly business. It should also be noted that some carmakers, particularly the European assemblers, have also taken over the distribution business from their long-time Thai partners because of the financial and other business difficulties the Thai partners were experiencing.

Between 1965 and 1985, most of the parts firms were still Thai owned, but

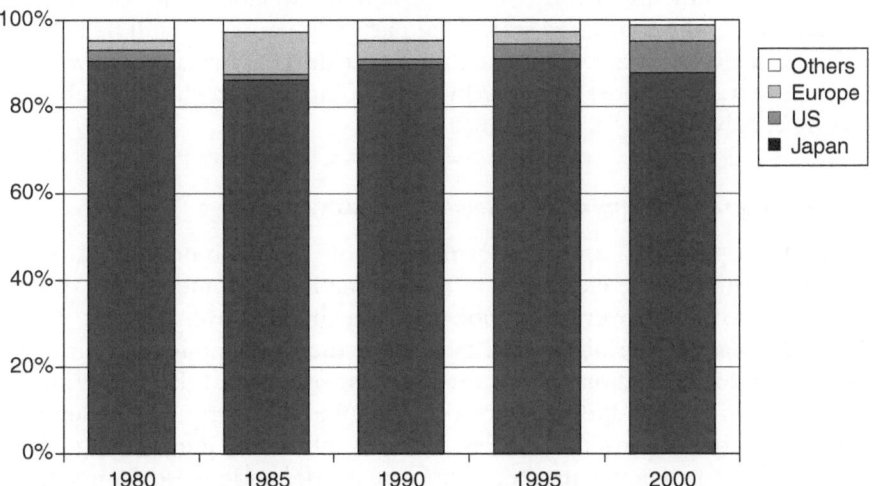

Figure 3.1 Market share in the automobile industry in Thailand

Source: Thai Automobile Industry Association

at least thirty Japanese component suppliers established their businesses during this period (Poapongsakorn 1997). Most of them were large OEM producers attracted to Thailand by the local content policy introduced in the early 1970s. The industrial boom in 1986–1991 saw thirty-seven new firms investing in Thailand. The liberalization in the early 1990s also attracted more technologically advanced, though smaller, parts suppliers to Thailand. At the same time, American component suppliers also followed their car assembly partners into Thailand, including Dana, Visteon and Delphi. Their presence has not only intensified the competition between parts suppliers, but also affected the status of Thai suppliers. Since the American suppliers have closer business relationships with the carmakers, particularly in R&D, they have replaced the Thai (and Japanese) firms as the first-tier suppliers. The Thai firms have to work for those first-tier American firms. Moreover, the American assemblers have also imposed a new standard, QS9000, on Thai parts producers wanting to supply them with parts and raw materials.

The economic crisis has also forced the takeover of many medium- and large-scale Thai parts suppliers by their Japanese partners. Interestingly, the crisis did not provoke a massive exodus of foreign auto and parts firms from Thailand. The worst cases were considerable downscaling (e.g. GM) and postponement of investment projects (e.g. Ford). In fact, new foreign parts suppliers have entered the market since 1998. Japanese foreign direct investment (FDI) increased from US$136 million in 1996 to $446 million in 1997 and $234 million in 1998, but declined to $105 million in 1999 (Farrell and Findlay 2001: 26). It should be noted, however, that most FDI in the 1998–1999 period was not 'green' investment (Table 3.5).

In effect, foreign investors now almost completely dominate the automotive industry. Almost all assembly firms are now 100 per cent foreign owned, while most of the small and medium-sized Thai parts suppliers have been marginalized, producing only the low value added and labour-intensive parts. At the same time, the four large-scale Thai parts companies are still heavily in debt. Fortunately, there are still some medium-scale Thai firms that have successfully penetrated export markets by developing their technological and marketing capability.

Factors affecting restructuring of the automobile industry

The structural changes and modernization of the automobile industry referred to earlier can be explained by a combination of domestic and external factors. Internally, government policies have changed over the past three decades. During the period 1970–1985, when the main policy was import substitution industrialization, government tax incentives and the high wall of tariff protection had helped create a fragmented industry. The assembly operations, each with several models, were highly inefficient, selling to a small but highly protected market. Their heavy dependence on imported parts and components also contributed to serious trade deficits. Despite rapid economic growth during this period, the industrial structure remained highly inefficient and fragmented.

Table 3.5 Net flows of foreign direct investment in machinery and transport equipment in Thailand in millions of US dollars

Year	Machinery and transport equipment	Industrial sector	Share in industrial sector (%)
1974	3	52	1.56
1976	0	23	0.00
1978	2	23	3.57
1980	5	49	2.65
1982	10	53	5.32
1984	5	135	1.21
1986	−1	81	−0.38
1987	6	186	1.69
1988	25	639	2.26
1989	43	852	2.42
1990	97	1,217	3.82
1991	90	935	4.43
1992	43	369	2.00
1993	62	452	3.58
1994	12	513	0.91
1995	145	567	7.24
1996	109	709	4.80
1997	396	1,820	10.92
1998	661	2,209	12.85
1999	394	1,268	11.06
2000	667	1,813	23.71

Source: Bank of Thailand

The industry began to experience major structural change in the early 1990s as a result of a liberalized policy for the automotive industry and the adoption of AFTA. In response to the high demand for cars, the government initiated a series of deregulation reforms and trade liberalization, which included reducing tariffs for imported completely built up (CBU) vehicles, lifting of capacity control and model restriction and liberalization of the taxi industry. These policies attracted both existing foreign automotive firms and new, non-Japanese carmakers and their parts suppliers to Thailand, resulting in a deepening of the industrial structure.

Externally, economic liberalization in many developing countries, particularly in Asia and Latin America, has led to progress in information technology, and rapid world economic growth since the late 1980s has triggered a series of strategic changes by auto multinational enterprises (MNEs). Faced with a saturated market in the developed countries, MNEs began to invest in some developing countries that had high growth potential and had liberalized their economies. The competition led to over-investment, which resulted in global overcapacity in the mid-1990s. The MNEs, therefore, were forced to adjust their strategies in order to gain a larger market share by seeking new markets and protecting their existing market share. The strategies they adopted include investing in new facilities and the modernization of old plants in countries with high growth potential, and/or where production costs are lower; reorganizing the global supply network in order to support lean production; and centralizing design and R&D in parent companies, exploiting information technology to reduce costs; and, finally, merging and making use of strategic alliances. In this new approach, the MNEs considered activities in some developing countries as integral parts of their global produc- tion strategies. Although tariffs and taxes are still high in most developing countries, including Thailand, FDI is no longer for the purpose of tariff jumping, as was the case in the 1960s and 1970s. Instead, the MNEs have planned to use developing countries, such as Thailand, not only as one of their markets, but also as a production base for exports (of both parts and cars) to neighbouring and regional markets so that the minimum efficient plant size can be reached. Gone are the days of sub-optimal-sized plants able to survive only because of high walls of tariff protection.

In sum, the industrial restructuring and modernization process in the automobile industry has taken the following forms (UNCTAD 2000: 167).

- The new plants have larger production runs in order to exploit the scale economies made possible by enlarged and liberalized regional markets, such as ASEAN and East Asia. Moreover, the new plants are more capital and technology intensive, and have adopted new employment practices.
- The pressure to enhance scale economies has led the MNEs to adopt the strategy of specialization and standardization among their affiliates in terms of models and component ranges. For example, Thailand is desig- nated as the regional production base for one-ton pick-up trucks by some MNEs. Standardization in design, development and production has led to growing centralization of R&D in developed countries and, correspondingly, less adaptive work in developing countries. Moreover, since the parts suppliers in developing countries have to be part of the MNE's global production network, they have to invest more in modern, high-technology machinery.
- Assemblers, particularly the American Big Three, prefer to develop collab- orative relationships with a limited number of core suppliers who assume responsibilities for modular manufacturing. These suppliers are often technologically advanced companies that have followed their automaker customers abroad. They take on the role of first-tier suppliers and push both the local and some Japanese parts suppliers to the second tier.
- The economic crisis of 1997–1998 urgently forced the car assemblers and parts suppliers to step up their exporting efforts to partially compen- sate for the sharp drop in domestic vehicle sales. Even Toyota, which did not have any export plans in 1997, decided to begin exporting pick-up trucks in late 1998 after it was severely hit by the crisis. As a consequence, exports of car and auto parts have overtaken imports for the first time.

Thai policies regarding technology transfer

The Thai government does not have a compulsory[5] but, rather, an 'open- door' policy towards technology transfer.[6] Technology transfer is generally regarded as a by-product of FDI promotion. Only after the economic crisis did the BOI set up an 'Industrial Human Resources Development Unit' to begin promotion of human resource development (HRD) and technology transfer. Although the BOI requires applicants for promotional privileges to have technical plans, it does not monitor the plans.

Busser (1999: 233–234) reports that all fully Thai-owned firms and Thai–Japanese joint ventures have technical agreements (formal technology transfer) with Japanese providers of technology. From interviews[7] with both some government officials and local auto parts firms by the author, it can be concluded that Thai government officials are not involved in the negotiations or enforcement of these technical agreements. It seems that the public sector has inadequate levels of technical, administrative and negotiating capacities[8] with regard to technical agreements. Thus it is difficult for the Thai government to intervene in formal technology transfer.

For auto assembly, Kanokwan (1996: 121–126) reports that the technology of Japanese and German auto assemblers in Thailand almost always originates from their parent companies. Intra-firm technology transfer is achieved through agreements involving royalty payments and some tied-in clauses, such as export restriction. Kanokwan also points out that, since government policy has ignored technology transfer, these technology transfers have not been very deep.

However, there are some major Thai policies, such as those applying to local content requirement, R&D promotion, export promotion and HRD, that have indirectly shaped technology transfer and the formation of domestic technological capability in the auto assembly and parts industries. I shall now discuss LCR and R&D promotion policies in detail.

Local content requirement (LCR)

The Thai government introduced the LCR policy in 1971 in order to increase local industrial capability and stimulate technology transfer to local firms (BOI 2001: 9). The LCR compelled auto assemblers to use a greater proportion of locally produced parts. As a result, the auto part industry in Thailand expanded substantially. Many auto parts firms, including Japanese and Thai firms, and joint ventures, were established after 1971. For example, Nippondenso (Thailand), Izumi Piston Manufacturing (Thailand) and Summit Auto Seat Industry were established in 1972. Doner (1991: 47) reports that Thai auto parts firms increased in number from several dozen in 1970 to over 200 by the middle of the 1980s. From 1971 onwards, the level of domestically produced parts under the LCR gradually increased from 20 per cent for passenger cars in 1974 to 54 per cent in 1990. For pick-up trucks, which are an important part of the Thai automobile market, a distinction was made between petrol- and diesel-fuelled vehicles. The LCR for petrol-fuelled vehicles was raised to 60 per cent in 1994, while the percentage for diesel-fuelled pick-ups was set at 72 per cent. In 2000, all LCR regulations were abolished. Table 3.6 shows LCRs in 1994 based on assigned points, each component group, such as the exhaust system or the suspension system, being awarded a set number of points.

Table 3.6a Local content requirements, 1994 (based on assigned points; see Table 3.6b)

Vehicle type	Points requirement
Passenger cars	54
Pick-up trucks	
Petrol	60
Diesel	72
Large trucks	45

Table 3.6b Assigned points for components of passenger cars

Name of component groups	Number of components	Assigned points
Base engine	31	15.3
Ancillary engine components	29	7
Electrical components	16	4
Wiring system	6	2
Exhaust system	4	2
Fuel system	7	2
General components	4	10
Panels and soft trim	24	4.25
Seats	5	5
Glass	5	2.5
Lamps	6	1
Suspension system	9	3.5
Brake system	27	3.1
Clutch	10	1.9
Body parts	41	23
Other body parts	5	1.45
Transmission	22	4.3
Steering system	11	2.95
Final drive	20	3.75
Instrument panel	12	1
Total	294	100

Source: BOI (1995)

To supply qualified auto parts, most local OEM partmakers have technical assistance (TA) agreements, a formal type of technology transfer, with Japanese companies. For example, in the Somboon Group, Somboon Malleable Iron Industrial (a 100 per cent Thai-owned company established in 1975) has TA agreements with Asahi Tech. Corporation (Japan) and Ibara Seiki (Japan). Bangkok Spring Industrial (a 100 per cent Thai-owned company established in 1976) has a TA agreement with Mitsubishi Steel Mfg (Japan). Somboon Advance Technology (a 100 per cent Thai-owned company established in 1995) has a TA agreement with Gohsyu Inc. (Japan) and Ibara Seiki (Japan).

On the other hand, Japanese auto assemblers have to transfer technology to local firms in order to get qualified auto parts at a low price. As regards inter-firm technology transfer, Kriengkrai (2002) reports that technology is being transferred from Japanese auto assemblers to local auto partmakers, including Thai firms. Most of these technology transfers are related to quality control, production management, design of moulds and dies and machinery maintenance. However, Maruhashi (1995) pointed out that, although the LCR increased local production capacity of auto parts in Thailand, for some auto parts the Japanese auto assemblers prefer in-house production to subcontracting.

In sum, the LCR policy forced auto assemblers, mostly Japanese auto assemblers, to increase production locally in Thailand. In order to get qualified auto parts at a low price, Japanese auto assemblers have to help local auto partmakers by either transferring technology directly or arranging TA agreements between local suppliers and Japanese auto partmakers.

R&D promotion

In order to promote technology transfer and the formation of domestic technological capabilities, the Thai government has, since the 1980s, employed fiscal and financial measures to promote R&D within the industrial sector (Nit *et al.* 1998: 19–45). However, the R&D promotions have been widely distributed to many fields and industries, only a few of which are in the automobile industry. It seems that the R&D promotion has had only a small effect on the formation of domestic technological capabilities in the automobile industry, a consequence of its limited resources, as discussed below.

1   The BOI has given R&D promotional privileges since 1989. For promotional purposes, it classifies R&D projects into three categories, as follows.[9]

   a   *R&D projects as a separate unit from other businesses.* For these R&D projects in zones 1 and 2,[10] the BOI grants two promotional privileges (Table 3.7). They include: (a) a 50 per cent reduction of import duties on machinery (which is not included in the Announcement of the Ministry of Finance no. 13/2533, 1990, and the import duty rate is not less than 10 per cent); and (b) corporate income tax exemption for a period of eight years.

   For R&D projects in zone 3, the BOI grants the most promotional privileges: (a) import duty exemption for machinery; (b) corporate income tax exemption for the first eight years, and 50 per cent reduction of corporate income tax for an additional five years; (c) a 200 per cent deduction for expenditure on transportation, electricity and water supply; and (d) a 25 per cent deduction for the cost of constructing infrastructure.

   Twenty-five R&D projects were promoted under this category in the period 1989–1994 and seventeen in the period 1995–2000. The average value of the investment project was small, roughly 99 million baht per project during 1995–2000. The total value of these projects

Table 3.7 BOI promotion of R&D projects (category 1: separate R&D)

	1995	1996	1997	1998	1999	2000
1 No. of R&D projects granted promotion and in operation	4	5	1	3	1	3
2 Value of investment (in millions of baht)	466.5	424.2	28	474	119	171
3 Gross domestic private investment (in billions of baht)	1,343.2	1,415.1	1,030.6	583.4	500.2	573.2
4 Proportion of 2 in 3	0.000347	0.0003	0.000027	0.000812	0.000238	0.000298

Source: Board of Investment

in proportion to gross domestic private investment is negligible (Table 3.7). Only a few of these projects are in the automobile industry. For example, the R&D projects at Thai Rung Union Car, a local auto partmaker and semi-assembler, and at Summit Auto Body Industry, another local auto partmaker, received the promotion in 1990 and 1999 respectively.

b   *R&D projects as an integral part of an investment project from its inception.* The BOI has granted promotional privileges to these projects, as it has for other businesses. In the automobile industry, Honda Cars Manufacturing Thailand received the promotional privileges for its R&D project at the Ayutthaya plant.

c   *R&D projects as an additional unit of existing promoted projects.* The R&D projects under this category receive some additional privileges over the existing promoted normal projects. The supplemental promoted privileges include an additional three-year exemption from corporate income tax and an exemption from import duties on R&D machinery.

2   The Ministry of Finance has given: (a) a depreciation allowance for R&D machinery and equipment; and (b) a tax concession for R&D expenditures since 1994. It has granted the 40 per cent allowed depreciation of the price of machinery and equipment used for R&D (calculated from date of acquisition). In addition, the taxable income of firms is allowed a 200 per cent deduction on R&D expenditures. However, this tax concession is only for the R&D expenditures paid out to R&D organizations that are on the approval lists of the Ministry of Finance.

Nit *et al.* (1998: 23–27) report that statistics on the utilization of depreciation allowances and tax concessions are not available; they also point out that the Ministry of Finance's definitions of R&D activities, and R&D machinery and equipment, are not clear. This unclear definition is one major obstacle to the utilization of depreciation allowances and tax concessions.

3   Several government agencies have provided soft loans and grants to support R&D projects in the private sector.

a   *Soft loans from the Ministry of Science, Technology and Environment (MOSTE).* Under MOSTE, the Research and Technology Development Revolving Fund has provided soft loans to support R&D projects in the private sector since 1987. It provided small soft loans to forty R&D projects, averaging 8 million baht per project during the period 1987–1997 (Nit *et al.* 1998: 28–29). Unfortunately, the automobile industry is not included among the industries eligible for this scheme. However, supporting industries, such as machinery, metal and materials, are eligible.

b   *Soft loans and grants from the National Science and Technology Development Agency (NSTDA).* The NSTDA has provided soft loans and grants to support R&D projects in the private sector since 1988. It gives priority to only three areas: (1) bioscience and biotechnology;

(2) materials technology; and (3) applied electronics and computers. Between 1988 and 1997, the NSTDA provided soft loans to thirty-one R&D projects that came to a total of 489 million baht, and grants to ten projects totalling 43.3 million baht.

c   *Grants from the Thailand Research Fund (TRF).* The TRF has provided grants to support both basic and applied research projects since 1994. However, it mainly supports research projects of public organizations. In the private sector, only six research projects received grants, amounting to a total of about 6.3 million baht, from the TRF during the period 1994–1997.

d   *Soft loans from the Bank of Thailand (BOT).* The BOT has provided soft loans to support R&D projects in the private sector since 1989. However, the effective interest rate and conditions of these loans are not very attractive to R&D projects. Thus only a few projects took out soft loans from the BOT.

Achievements and problems concerning technology transfer

In the Thai case, there is some evidence suggesting successful technology transfer in the automobile industry.

1   *Development of local auto parts and supporting industries.* In Thailand, the local auto parts industry and local supporting industries have to rely on the technology of auto assemblers and transnational (either global or regional) auto partmakers. To some extent, their success may indicate the achievement of technology transfers from automobile assemblers and transnational auto partmakers.

Under the LCR policy, the high import duty on CBU cars, and the lower import duty on completely knocked down (CKD) parts, the local auto parts industry has been developing since the 1970s. Doner (1991: 47) reported that the number of Thai auto parts firms increased from several dozen in 1970 to over 200 by the middle of the 1980s; 150 of these were OEMs. This industry then grew rapidly during the 1985–1996 period, owing to the appreciation of the yen, the increase in domestic demand for cars and the LCR policy. In terms of the number of auto partmakers in Thailand, Higashi (1995: 20) estimated that there were about 150 OEM partmakers and 200–250 REM partmakers in 1995. However, BOI (1999c: 2) reported that Thailand had about 200 OEM partmakers and about 500 REM partmakers in 1999.[11]

In terms of the technology level of the supporting industry in Thailand, of auto parts and electrical and electronic partmakers, JICA and the Ministry of Industry (1995) ranked it at the upper middle level overall of technological capability in ASEAN. Specifically, the die-casting (aluminium alloy) sector makes use of a high level of technology, just slightly lower than the average technology level found in industrial countries. The technology utilized in the die-forging (steel) and iron and steel casting sectors are ranked at the upper to top levels in ASEAN. The levels of technology used in presswork and rubber processing in Thailand are ranked at the upper middle level in ASEAN.

2   *Total factor productivity in the auto assembly sector.* The growth of total factor productivity (TFP) represents technical progress. Thus the growth of TFP may be partly due to the achievement of technology transfer, or to the formation of domestic technological capability.

On the basis of the growth accounting framework, it was found that the average annual rate of TFP of auto assemblers (TSIC 38431) increased from 4.8 per cent in 1982–1990 to 8.8 per cent in 1990–1994 (Table 3.8). This indicates that the auto assembly industry in Thailand has experienced technical progress. This technical progress partly represents a formation of human capital through training and intra-firm technology transfer.

3   *Domestic industrial linkage of the automobile industry.* The degree of domestic industrial linkage can indicate the capability of local supporting industry. It may also indicate the achievement of technology transfer from auto assemblers to the local auto parts industry and supporting industry in Thailand. In Thailand, this industry was first limited to assembling knocked-down imported cars. After the LCR was imposed, the local auto parts and supporting industries grew substantially. As a result, domestic backward linkages have increased significantly. Table 3.9 shows that the total backward linkage index in the automobile industry increased from 1.46 in 1985 to 1.69 in 1998.

In addition, according to the Input–Output Table with 180 sectors of the National Economic and Social Development Board, the local content in the motor vehicle sector (Sector 125) increased from 39 per cent of input value in 1985 to 41.9 per cent in 1998.

Table 3.8 Annual growth of total factor productivity (TFP)

Assembly of automobiles (TSIC 38431)	1982–1990 (%)	1990–1994 (%)
Value added	18.2	41.0
Factor input		
Labour	20.7	25.1
Capital	3.9	−1.8
Total factor productivity (TFP)	4.8	8.8

Notes
This was calculated using data from the report of the industrial survey (whole kingdom) conducted by the National Statistical Office of Thailand. The figures represent the total of both small-scale establishments (with 10–19 persons) and large-scale establishments (with more than 20 persons).

Value added data were deflated by the wholesale price index for transportation equipment. Labour input has been measured as the total wages and salaries paid and deflated by consumer price index. Capital input is obtained from the book value of fixed assets, which refers to the net value of fixed assets after deducting the accumulated depreciation at the end of the year. Capital input data were deflated by capital price index.

The theoretical framework for this calculation is based on the growth accounting framework. See more details of the theoretical framework in Tinakorn and Sussangkarn (1998)

Table 3.9 Domestic backward linkage (24 × 24 sectors)

Sector	Direct			Total			Rank[a]
	1985	1995	1998	1985	1995	1998	
Agriculture	0.28	0.27	0.30	1.44	1.41	1.47	—
Mining	0.26	0.30	0.29	1.37	1.42	1.43	—
Food processing and animal feed	0.62	0.60	0.60	1.94	1.89	1.93	3
Textiles and garment industry	0.51	0.52	0.53	1.90	1.92	1.95	2
Automotive and parts industry	0.29	0.32	0.42	1.46	1.48	1.69	7
Electronics and electrical appliances	0.41	0.22	0.30	1.63	1.32	1.45	13
Ceramics and glass industry	0.48	0.45	0.48	1.70	1.66	1.74	5
Rubber and rubber products industry	0.72	0.70	0.64	2.09	2.05	2.01	1
Shoes and leather products industry	0.56	0.39	0.35	2.00	1.64	1.56	12
Pharmaceutical and chemical industry	0.39	0.43	0.49	1.61	1.66	1.82	4
Jewellery	0.27	0.40	0.43	1.38	1.62	1.68	8
Wood and wood products	0.43	0.36	0.41	1.65	1.54	1.64	10
Plastic products	0.31	0.44	0.43	1.49	1.70	1.73	6
Petrochemical industry	0.18	0.07	0.10	1.27	1.10	1.14	15
Iron and steel	0.48	0.22	0.29	1.75	1.32	1.43	14
Paper and publishing industry	0.25	0.31	0.40	1.37	1.45	1.63	11
Industrial machinery manufacturing	0.36	0.36	0.43	1.55	1.51	1.67	9
Other manufacturing	0.47	0.43	0.43	1.74	1.64	1.68	8
Public utilities	0.49	0.46	0.57	1.75	1.67	1.87	—
Construction	0.55	0.45	0.50	1.87	1.68	1.79	—
Trade	0.19	0.22	0.22	1.28	1.33	1.34	—
Services	0.26	0.28	0.28	1.43	1.45	1.46	—
Transportation	0.37	0.37	0.51	1.52	1.55	1.78	—
Other	0.94	0.89	0.37	2.58	2.41	1.59	—

Source: Terdudomtham (2001)

Note
These figures are calculated from the Input–Output Table of Thailand 1985, 1995, 1998 (Input Value of Domestic Products), Office of the National Economic and Social Development Board. The theoretical framework is based on Miller and Blair (1985)

a The ranking is based on total backward linkage of fifteen industries in 1998

4   *Some problems of technology transfer.*[12] Currently, local auto partmakers need higher product design capabilities to satisfy the demands of auto assemblers. However, they generally lack testing facilities for designing and have to face the high costs of software and machines needed for designing. Moreover, under the system by which auto assemblers order auto parts as modules, it is meaningless for local auto partmakers to design their own individual parts.

Many local auto partmakers receive technology through TA agreements. For some types of advanced technology, the TA fee is very high, possibly up to 8 per cent of sale revenue. It is worth noting that the Thai government has never been involved in the negotiations of TA agreements.

Concluding remarks

1 The Thai government does not have a compulsory policy for technology transfer. It has an 'open door' policy towards technology transfer. Technology transfer is generally regarded as a by-product of FDI promotion. This is partly because the Thai government is inadequately aware of the technology transfer issue. In addition, the public sector has inadequate levels of technical, administrative and negotiating capacities to be involved in the negotiations of technical agreements in private firms.

2 However, there are some major Thai policies that have indirectly stimulated technology transfer and the formation of domestic technological capabilities in the auto assembly and parts industries, such as the LCR and R&D promotion.

The LCR policy forced auto assemblers, mostly Japanese auto assemblers, to increase production locally. In order to obtain qualified auto parts at a low price, Japanese auto assemblers had to help local auto partmakers by either transferring technology directly or arranging TA agreements between local suppliers with Japanese auto partmakers. In other words, the LCR was the most important policy that indirectly forced intra-firm technology transfer from Japanese auto assemblers to local auto partmakers through a subcontracting system. In addition, under the LCR, auto assemblers increased their in-house production of auto parts, and intra-firm technology transfer also increased through the training of their workers.

In order to promote the formation of domestic technological capabilities, the Thai government has employed fiscal and financial measures to promote the R&D in the industrial sector since the 1980s. The R&D promotions are distributed widely to many fields and industries, only a few of which are in the automobile industry. It can be concluded that the R&D promotion has had only a small effect on the formation of domestic technological capability in the automobile industry, because of its limited resources.

3 In Thailand, there is some evidence that indicates the achievement of technology transfer in the automobile industry. First, local auto parts and supporting industries have been developed substantially in terms of the number of firms and their production. Their overall technological development is ranked at the upper middle level within ASEAN. Second, there is the growth of TFP in this sector, especially in auto assembly (TSIC 38431). Third, domestic industrial linkages in the automobile industry, especially backward linkages, have increased over time. The ratio of domestic input in the automobile industry has also increased over time. However, there are some problems for technology transfer within the automobile industry. For example, the TA fee is relatively high.

4   Since the 1997 crisis, the BOI has paid more attention to technology transfer. It set up the Industrial Human Resource Development Unit to begin the promotion of HRD and technology transfer after the crisis. In addition, with the cooperation of the private sector, the Thai government set up the Thailand Automobile Institute (TAI) in 1998 in order to promote the development of the automobile industry, especially the auto parts industry. With the cooperation of the Japan Automobile Manufacturers Association, the Japan Automotive Parts Industry Association and the Japan Automobile Research Institute, the TAI has provided TA to local auto partmakers in the areas of casting, forging, pressing, machining, product designing and so on. In sum, the Thai government has paid increased attention to technology transfer since the 1997 crisis.

Notes

1 See more discussions in Thamavit *et al.* (2000).
2 See more discussions in Peera (2001).
3 Prayoon Shiowattana (1991: 175–176) defined technology transfer as 'a learning process wherein technological knowledge is continually accumulated into human resources that are engaged in production activities'. The learning process is divided into four stages: acquisition, operation, adaptation and innovation.
4 Because of heavy debt, the Siam Cement Group decided to sell off most of its non-core business activities, including those in the automotive business. The pull-out reflects a change in its corporate strategy, because the SC group no longer believes that it can be the competitive leader in this industry.
5 Based on the discussions with Salil Wisalswadi from the BOI, and Pichai Tangchanachaianan and Nattapol Rangsitpol from the Office of Industrial Economics, Ministry of Industry.
6 ESCAP (1984: 19–24) pointed out that Thailand had followed an open-door policy towards international technology transfer, and Thailand's large payments for TA fees were associated with foreign investment.
7 Interviewed by the author.
8 ESCAP (1984: 17–26) reported that the degree of government intervention in technology imports in the Asian-Pacific countries varied from country to country. It depended on the awareness and seriousness of the host government, and the technical, administrative and negotiating capacities of the private and public sectors in the host country.
9 See more discussions in Yuthasak (1997: 89–99).
10 Based on the BOI classification, zone 1 includes Bangkok, Samut Prakan, Samut Sakorn, Pathumtani, Nonthaburi and Nakornpathom. Zone 2 includes Samut Songkram, Ratchaburi, Kanchanaburi, Suphanburi, Angthong, Pranakorn Sri Ayuthya, Lopburi, Nakornnayok, Chachoengsao and Chonburi. Zone 3 includes all other provinces and Laem Chabang Industrial Estate (in Chonburi).
11 While BOI (1999b: 2) reported that there were about 700 auto partmakers in Thailand, BOI (1999a: 3) reported that there were about 600 auto partmakers in the country, of which about 150 were in the OEM market.
12 The author thanks Yongkiat Kitaphanich (Vice President of Somboon Group), Pichid Uasakunkiat (Vice President of Bangkok Pacific Steel) and Somsak Pongsakornvanich (Bangkok Pacific Steel) for their informative discussions.

References and further reading

Board of Investment (BOI), Thailand (1995) *Investment Opportunities Study: Automotive and Auto Parts Industries in Thailand*, Bangkok.

—— (1999a) *Automotive Industry Development in Thailand* (May), Bangkok.

—— (1999b) *The Executive Report: Thailand's Automotive Parts Industry*, Bangkok: Sukjai Management Services.

—— (1999c) *The Executive Report: Thailand's Supporting Industry*, Bangkok: Sukjai Management Services.

—— (2001) *Automotive Industry in Thailand: Development and Current Status*, Bangkok.

Brimble, P. (2000) 'Mergers and acquisitions in Thailand', paper given at the seminar 'Cross-border M&As and Sustained Competitiveness in Asia: Trends, Impacts and Policy Implications', 9–10 March, Bangkok: UNCTAD and Ministry of Foreign Affairs.

Brooker Group (1997) *Automotive Industry Export Promotion Project, Thailand Industry Overview, Final Report*, prepared for the Office of Industrial Economics, Ministry of Industry, Royal Thai Government (January), Bangkok.

Busser, R.B.P.M. (1999) 'Changes in Organization and Behavior of Japanese Enterprises in Thailand: Japanese Direct Investment and the Formation of Networks in the Automotive and Electronics Industry', unpublished Ph.D. dissertation, Leiden University, Leiden, the Netherlands.

Doner, F.R. (1991) *Driving a Bargain: Automobile Industrialization and Japanese Firms in Southeast Asia*, Berkeley: University of California Press.

ESCAP (1984) *Costs and Conditions of Technology Transfer through Transnational Corporations*, Bangkok: United Nations.

Farrell, R. and Findlay, C. (2000) 'Japan and the ASEAN Automotive Industry: Developments and Inter-relationships in the Regional Automotive Industry', paper given at a conference organized by the Australian–Japan Research Center, Asia Pacific School of Economics and Management, Australian National University, Canberra.

Fujimoto, T. and Orihashi, S. (2003) 'The Strategic Effects of Firm Sizes and Dynamic Capabilities on Overseas Operations: A Case-Based Comparison of Toyota and Mitsubishi in Thailand and Australia', in Busser, R. and Sadoi, Y. (eds) *Production Networks in Asia and Europe: Skill Formation and Technology Transfer in the Automobile Industry*, London: Routledge.

Hayashi, T. (1990) *The Japanese Experience in Technology: From Transfer to Self-Reliance*, Tokyo: United Nations University Press.

Higashi, S. (1995) 'The Automotive Industry in Thailand: From Protective Promotion to Liberalization', in Institute of Developing Economies (ed.) *The Automotive Industry in Asia: The Great Leap Forward?*, Tokyo: Institute of Developing Economies.

Japan International Cooperation Agency (JICA) and Thailand, Ministry of Industry (1995) *The Study on Industrial Sector Development: Supporting Industries in the Kingdom of Thailand*, Tokyo: Unico International Corporation.

—— (1995) *The Follow-up Study on Supporting Industries Development in the Kingdom of Thailand*, Tokyo: Unico International Corporation.

Kanokwan Booksabokkeaw (1996) 'Technology Transfer in the Automobile Assembly: A Case Study of Japanese and German Firms', unpublished MA thesis, Chulalongkorn University (in Thai).

Kriengkrai Techakanont (2002) 'A Case Study of Inter-firm Technology Transfer in Automobile Industry of Thailand', *Thammasat Economic Journal*, 20 (1) (March): 12–46 (in Thai).

Maruhashi, H. (1995) 'Japanese Subcontracting System in Thailand: A Case Study of the Thai Automobile Industry', unpublished Master's thesis, Faculty of Economics, Thammasat University, Bangkok.

Miller, R.E. and Blair, P.E. (1985) *Input–Output Analysis: Foundations and Extensions*, Englewood Cliffs, NJ: Prentice-Hall.

Ministry of Industry, Office of Industrial Economics (2001) 'Automotive Industry in Thailand' (January), Bangkok.

Nit, Chantramonklasri; Tinakorn, Pranee; Tangkitvonich, Somkiat and Virasa, Thanaphol (1998) *Effective Mechanisms for Supporting Private Sector Technology Development and Needs for Establishing Technology Development Financing Corporation*, Bangkok: Thailand Development Research Institute Foundation.

Peera Charoenporn (2001) 'Automotive Part Procurement System in Thailand: A Comparison of American and Japanese Companies', unpublished Master's thesis, Faculty of Economics, Thammasat University, Bangkok.

Poapongsakorn, N. (1997) 'ASEAN Automobile Industry in the Emerging Economic Integration Environment: Corporate Strategies and Government Policies', paper prepared for the Second Expert Meeting of the Interregional Project 'Transnational Corporations and Industrial Restructuring in Developing Countries' organized by Division of Investment, Technology and Enterprise Development, Geneva: UNCTAD, November.

Prayoon Shiowattana (1991) 'Technology Transfer in Thailand's Electronics Industry', in Yamashita, S. (ed.) *Transfer of Japanese Technology and Management to the ASEAN Countries*, Tokyo: University of Tokyo Press.

Terdudomtham, T. (1997) 'The Automobile Industry in Thailand' (ASP-5 Subprogramme on Liberalization on Trade and Investment), Bangkok: Thailand Development Research Institute.

—— (2001) 'Industrial Linkage in Thailand: 1985–1998', a paper (in Thai) for the UNIDO Integrated Programme for Thailand – Component 6 'Tracking Manufacturing Performance: Towards an Early Warning Mechanism Geared to the Real Economy'.

Thamavit, T., Kriengkrai T. and Peera, C. (2000) 'The Changes in Automobile Industry in Thailand', paper for the Nineteenth Annual HOSEI University International Conference 'Japanese Foreign Direct Investment and Structural Change in the East Asian Industrial System: Global Restructuring for the Twenty-first Century', organized by Hosei University, 30 October and 1 November, Tokyo.

Tinakorn, P. and Sussangkarn, C. (1998) 'Total Factor Productivity Growth in Thailand: 1980–1995', in *Competitiveness and Sustainable Economic Recovery in Thailand*, Bangkok: NESDB and the World Bank Thailand Office.

United Nations Conference on Trade and Development (UNCTAD) (2000) *The Competitiveness Challenge: Transnational Corporations and Industrial Restructuring in Developing Countries*, Geneva: United Nations.

Yuthasak Kanasawadi (1997) 'BOI and Promotion for Research and Development', *Thailand's Investment Promotion Journal*, 8 (3) (March): 89–99.

Personal communications

(2001) Interview with Salil Wisalswadi from the BOI by the author.

(2001) Interview with Pichai Tangchanachaianan and Nattapol Rangsitpol from the Office of Industrial Economics, Ministry of Industry by the author.

(2001) Interviews with government officials and private-sector representatives by the author.

(2001) Interview with Yongkiat Kitaphanich, Vice President of the Somboon Group, by the author.

(2001) Interview with Pichid Uasakunkiat, Vice President of Bangkok Pacific Steel, by the author.

(2001) Interview with Somsak Pongsakornvanich, Bangkok Pacific Steel, by the author.

4 Malaysian policies for the automobile sector

Focus on technology transfer

Tham Siew-Yean

The manufacturing sector in Malaysia was developed with the help of both 'market-friendly' policies and direct state intervention. This dual approach can be attributed to the twin objectives of the New Economic Policy (NEP) that was promulgated in 1970 to direct economic development in the country. The NEP has as its twin objectives both growth and equity. While selective 'open' policies, such as export and promotion of foreign direct investment (FDI), were utilized to promote growth, the redistribution goal led to considerable state intervention in the distribution of wealth, employment, equity ownership and the development of specific sub-sectors, such as the automotive sector.

To illustrate, by 1990 the Malaysian state was targeting the allocation of corporate wealth and employment accrual to indigenous Malaysians (Bumiputeras), other Malaysians and overseas investors by 30, 40 and 30 per cent respectively in order to redress the economic imbalances between the different ethnic groups in the country. State intervention in the development of the automotive sector can also be traced back to the redistribution objective, when dissatisfaction with the slow progress made by Bumiputeras' participation in the economy prompted the development of heavy industries (HI) and the launch of the HI programme in 1981. Weak linkages between the domestic economy and the Free Trade Zones also contributed to the shift in policy directions at that time. Consequently, the Heavy Industry Corporation of Malaysia (HICOM) was incorporated to oversee the production of cement and steel via its subsidiaries Perwaja Terengganu and Kedah Cement, respectively.

In the case of automobiles, the first National Car Project (NCP) was launched under HICOM's umbrella in an attempt to rationalize the local automotive industry and to foster growth in the rest of the industrial sector through technical spin-offs and linkage effects. At the same time, the NCP was targeted to assist and accelerate the participation of Bumiputeras in the automotive industry. Consequently, Proton (an acronym for Perusahaan Otomobil Nasional, or the National Automobile Enterprise) was established as a joint venture between Mitsubishi Motor Corporation (MMC), Mitsubishi Corporation (15 per cent equity each) and HICOM (70 per cent equity) in 1983. However, Proton was not the sole NCP in Malaysia. In fact, in the eighteen years since Proton, two other NCPs have been launched, each with a different technology partner. For example, the second NCP or Perodua, was launched in 1993 to produce mini passenger cars with Daihatsu (a subsidiary of Toyota Corporation), while a national truck project was set up in 1997 (with the assistance of Isuzu Motors). Concurrently, Proton, in its search for technological independence, had also forged new partnerships with Citroën and the Lotus Group International Ltd to develop new car models.

Given that Proton was set up in the early 1980s, it is now timely to assess the development of this sector, particularly in view of the rapid changes in the external environment and their potential impact on its future development. Because the auto industry is a policy-driven sector, it is especially important to determine whether government policies have been able to facilitate the transfer of technology necessary for this sector to be able to compete independently in an increasingly globalized world. The first objective of this chapter, thus, is to assess the impact of Malaysia's automotive policies on the development of this sector. In particular, the chapter examines the impact of these policies over time on the capacity for local production. Second, this chapter offers an analysis of the policy options available to Malaysia in view of increasing pressures to liberalize and globalize the world automobile industry.

The chapter is divided up as follows: the introduction is followed by a section that discusses the policies used to develop the automotive sector. The second section continues with an analysis of the impact of these policies on the overall development of the automotive sector, including technology development. This is followed by a review in the third section of the current constraints that can affect the future development of the automotive sector in Malaysia. The final section outlines the policy implications of these constraints. The chapter concludes with a summary of the main findings of this chapter.

Malaysian automotive policies

HICOM was established in 1980 with state capital (Chee 1992: 8), while Proton, as one of the companies set up by HICOM, was incorporated in 1983 as a joint venture between Malaysia and private investors in Japan, as explained in the introduction. State intervention for HICOM projects included the provision of subsidized loans, and the state also acted as guarantor of foreign loans. The automotive sector was protected from foreign competition by high tariff and non-tariff barriers. Table 4.1 shows the import duty structure of this sector in nominal terms, while Alavi (1996: 174) estimated the effective rate of protection of the transport and equipment subsector to be as high as 252 per cent in 1987. In contrast, Proton was given a preferential import duty rate of 13 per cent on completely knocked down (CKD) parts and a 50 per cent exemption from the excise duty (Mahani 1997: 32). In the case of Perodua, no import duty was provided and certain models were given partial exemption from excise duties. Non-tariff barriers, such as approval permits, were also used to curb imports for completely built up (CBU) units. In addition, the Ministry of International Trade and Industry (MITI) regulated the selling price of each model, as MITI's approval was required for pricing. The establishment of direct local marketing outlets

Table 4.1 Import duty structure in Malaysia

	Engine capacity	*CBU rate (%)*	*CKD*
Passenger cars	Less than 1,800 cc	140	42
	1,800–1,999 cc	170	42
	2,000–2,499 cc	200	60
	2,500–2,999 cc	250	70
	3,000 cc and above	300	80
4WD and MPV	Less than 1,800 cc	60	10
	1,800–1,999 cc	80	20
	2,000–2,499 cc	150	30
	2,500–2,999 cc	180	40
	3,000 cc and above	200	40
Others	Less than 1,800 cc	60	10
	1,800–1,999 cc	80	20
	2,000–2,499 cc	150	30
	2,500–2,999 cc	180	40
	3,000 cc and above	200	40
Trucks	All range	50	Nil
Buses	All range	50	Nil

Source: MIDA (2001: 107)

Notes
Excise duty is imposed on CKD cars only
First MYR 7,000: 25%
Next MYR 3,000: 30%
Next MYR 7,000: 50%
Next MYR 5,000: 60%
Balance exceeding MYR 25,000: 65%

further enhanced the development of the NCPs by addressing their distribution needs.

However, the automotive policies of Malaysia did not just focus on the manufacture and assembly of passenger cars. Rather, specific policies were designed to develop the local capacity for manufacturing automotive parts and components, namely, the mandatory deletion programme, the vendor development programme and the local material content policy.

Mandatory deletion programme

The Mandatory Deletion Programme (MDP) was introduced in 1980, even before the formation of HICOM. Under this programme, which is currently under review, it became mandatory that thirty specified components were to be produced locally. Accordingly, assemblers and manufacturers have to 'delete', or omit, the specified components from imported CKD packs and replace them with local procurements.

Vendor development programme

The First Industrial Master Plan (FIMP: 1986–1995), which was launched in 1986, utilized a targeted approach to accelerate growth of the manufacturing sector. The transport equipment industry was one of the twelve targeted sectors that was identified for development in line with the selective develop-ment of heavy industries as advocated by the plan. The development of this sector focused on the development of the NCPs and an effective local content programme in order to facilitate the development of linkages in the country (Anuwar 1992: 50). Consequently, the Vendor Development Pro-gramme (VDP) was launched in 1988 under the coordination of MITI to accelerate the development of local parts and components.

Under this programme, vendors are given technical support and guidance for their products by the VDP anchor, Proton, so that they can meet the stand-ards required by the technology provider. Thus Proton not only identifies technology partners for the vendors but, in some instances, has had to extend its own personnel to assist with the vendors' operations (Mahani 1997: 35). Table 4.2 indicates the technology assistance provided by Proton through the 'match-making' arrangement between the local vendors and reputable over-seas technical collaborators. As of 2001, there are 216 such 'matchmaking' arrangements, of which more than 50 per cent are sourced from Japan, owing to Proton's partnership with Mitsubishi and its supplier network. In turn, Proton received MYR 7 million under the Fifth Malaysia Plan (1986–1990) and an additional MYR 15 million under the Sixth Malaysia Plan (1991–1995) from MITI for its efforts in developing the VDP (Singh 2000a: 15).

Initially, a single sourcing system was utilized to guarantee and protect the market for its vendors as well as to enable the vendors to reach the necessary economies of scale in production. This also enabled the vendors to generate appropriate positive cash flows during the initial phase of their operations. However, Proton has unofficially moved towards multi-sourcing since 1996, although this policy was only made official in the year 2000. Under the multi-sourcing policy, up to four domestic suppliers can be tapped to supply each component group in an effort to inject competitive forces within the vendor supply system.

Since the vendors created are mainly small and medium-sized enterprises (SMEs), the VDP in the automotive sector also receives assistance from the Small and Medium Industries Corporation (SMIDEC), which was established in 1996 to promote and coordinate SME development in Malaysia. SMIDEC-

Table 4.2 Proton's 'matchmaking' collaborative arrangement

Source country	Joint venture	Technical assistance	Purchase agreement	Wholly owned company	Total
Japan	33	77	2	13	123
Germany	3	8	1	2	14
Taiwan	4	9	1	—	14
Korea	6	8	—	—	14
France	—	10	1	—	11
Others	10	19	2	7	38
Total	56	131	7	22	216

Source: Proton (2001: 12)

administered programmes include financing, providing information, developing human resources, fostering industrial linkages and assisting in technology acquisition. Thus in the case of financing, soft loans, guarantees on commercial loans, export credits and venture capital funding are provided for servicing the financial needs of the SMEs in all sectors. Furthermore, SMEs in all sectors have recourse to special funds such as the Technology Acquisition Fund (TAF) and the Industrial Technical Assistance Fund (ITAF) to upgrade their technology and to improve themselves in product development and design, quality, productivity, as well as market development.

Local Material Content Policy

The Local Material Content Policy (LMCP), first instituted in 1996, is applicable to both NCPs and assemblers (Table 4.3). While the VDP is directed specifically at the NCPs to assist the development of Bumiputera entrepreneurs in these projects, the LMCP is targeted at the assemblers who are franchise holders of global manufacturers (Singh 2000a: 12). The deadline was extended to 2000 and it is currently under review. Assemblers who fail to comply with the set targets have been warned of potential penalties in the form of price increments and import restrictions.

The Second Industrial Master Plan (SIMP), covering the years 1996–2005, reiterated the focus on the development of locally produced components while, at the same time, the need to deepen the development of this sector was also emphasized. Thus the plan noted the need to expand the scale of production and to upgrade technological capability, the development of human resources, product and engineering capability and production process technology (Malaysia 1996: 255).

Impact on overall development of the automotive sector in Malaysia

In view of the focus of government policies thus far, this study will highlight mainly the final assembly and parts and component industries in its

Table 4.3 Local content programme for motor vehicles

Vehicle type	Local content target (%)				
	1992	*1993*	*1994*	*1995*	*1996*
Passenger vehicles up to 1,850 cc	30.0	40.0	50.0	55.0	60.0
Passenger vehicles 1,851 cc to 2,850 cc Commercial vehicles up to 2,500 GVW	20.0	30.0	35.0	40.0	45.0
Passenger vehicles above 2,850 cc Commercial vehicles above 2,500 GVW	Mandatory deleted items only				

Source: MIDA (2001: 8)

Note
GVW, gross vehicle weight

assessment of the impact of the policies outlined above. In particular, it will focus on Proton's development, given the relatively longer period of government assistance that has been enjoyed by this company.

Automotive manufacturers and assemblers

The introduction of the NCPs changed the structure of the domestic automotive industry tremendously. Prior to the NCPs, assembly activities were fragmented and inefficient, with thirteen assembly plants producing a large number of makes and models for the domestic market (Mahani 1997: 10). With the arrival of the NCPs, Malaysian manufacturers came on-stream and the assemblers that used to dominate this sector became increasingly marginalized in terms of production and sales, leaving them either to focus on the upper segment of the market or to shift to component manufacturing.

Currently, the automotive assembly branch consists of four manufacturers and eleven assemblers (Table 4.4). The total production capacity of these fifteen manufacturers and assemblers amounts to 570,000 units of passenger and commercial vehicles per year (MIDA 2001: 79). Actual production of passenger vehicles initially dropped from 81,065 units in 1980 to 69,769 units in 1985, when the first Proton car made its debut, owing to the recession in that year (Table 4.5).

Subsequent recovery of the economy in the second half of the 1980s led to an increase in demand and production of Proton cars such that by 1990 Proton produced 73 per cent of the total output of passenger vehicles for that year. In 1996, just before the advent of the financial crisis in 1997, total passenger vehicle production increased to 280,944 units, with Proton contributing 64 per cent and Perodua, which made its debut in 1994, producing 17 per cent of this total. The attending recession in 1998 resulted in a sharp fall in production as demand plummeted. While Proton's share of total production fell marginally to 63 per cent in 1998, Perodua's share increased to 30 per cent as the share of production of non-Malaysian cars fell significantly from 18 per cent in 1996 to a mere 6 per cent the same year. Since the recovery of the economy in 1999, total production has climbed again, with Proton maintaining its leading share. It can also be observed from Table 4.5 that the improvement in production in 1999 was the largest for non-Malaysian cars.

As in the case of production, Proton has captured an increasing share of the market for passenger vehicles. As shown in Table 4.6, the market share taken by Malaysian cars increased steadily from 64 per cent in 1991 to 93 per cent in 2000. Proton's share peaked in 1994 (71 per cent), but the entry of Perodua that year resulted in a fall in this share to between 63 and 65 per cent from 1995 to 2000. During this same period, the share of Perodua increased progressively to reach 29 per cent in 2000. Conversely, the share of non-Malaysian cars fell successively to 7.3 per cent in the same year. Furthermore, this loss in the non-national car sector was more acute for Japanese cars, while the decline in the share taken by European and American cars was more gradual (Singh 2000a: 7).

Since the automotive sector is highly protected and local production of

Table 4.4 Automotive manufacturers and assemblers in Malaysia

Manufacturer/assembler	Category/makes	
	Passenger cars	*Commercial vehicles*
1 Perusahaan Otomobil Nasional (Proton)	Proton	
2 Perodua Manufacturing Sdn. Bhd.	Kancil, Daihatsu	Rusa, Kembara, Daihatsu
3 Associated Motor Industries Sdn. Bhd.	Ford, BMW, Mazda, Proton	Ford, Rover, Suzuki, Scania, Tata, Mahindra
4 Assembly Services Sdn. Bhd.	Toyota	Toyota, Hino
5 Asia Automobile Industries Sdn. Bhd.	Mercedes	Mercedes, Mazda Kia
6 Swedish Motors Assemblies Sdn. Bhd.	Volvo	Volvo, Suzuki
7 Tan Chong Motor Assemblies Sdn. Bhd.	Nissan, Audi, Peugeot	Nissan, Subaru
8 Oriental Assemblers Sdn. Bhd.	Honda, Peugeot, Mercedes	
9 Automotive Manufacturer (M) Sdn. Bhd.	Citroën, Proton	
10 Industri Otomotif Komersial (M) Sdn. Bhd.		Permas
11 Kinabalu Motor Assembly Sdn. Bhd.		Isuzu
12 Malaysian Truck and Bus Sdn. Bhd.		Isuzu, Mitsubishi, Iveco, MTB Perkasa, Pinzgauer, Ssang Yong
13 Bufori Motor Car Co. (M) Sdn. Bhd.	Bufori	
14 TVR Sports (M) Sdn. Bhd.	TVR	
15 TD Cars (M) Sdn. Bhd.	TD Cars	

Source: MIDA (2001: 99–101)

passenger vehicles is, as explained above, dominated by Malaysian cars that are geared primarily for domestic consumption, exports are quite small in number (Table 4.7); they mainly comprise passenger car exports from Proton (Singh 2000a: 9). As in the case of production and domestic sales, the crisis in 1997 also led to a fall in exports in 1998, a fall that experienced no recovery during 1999 and 2000. As shown in Table 4.7, exports fell successively, both in unit and in value terms, from 26,948 units or MYR 559.8 million in 1997 to 14,303 units or MYR 252.6 million in 2000.

A comparison of Table 4.7 with Table 4.8 reveals that Malaysia was a net importer of passenger cars for the entire period shown. This is due, in part, to the consumer preference for certain imported models and the focus of Malaysian cars on engines with a relatively small engine capacity, namely less than 1,500 cc. Imports of completely knocked down (CKD) kits far exceed those of CBU units, partially as a result of the tariff structure as well as the import dependence of the motor assemblers and manufacturers. Therefore, as shown in Table 4.8, CKD kits constitute more than 90 per cent of

Table 4.5 Production of passenger vehicles in Malaysia, 1980–2000 (in units)

Year	Total production	Proton	Perodua	Others
1980	81,065	—	—	—
1985	69,769 (100)	8,607 (12)	—	61,162 (88)
1990	116,526 (100)	85,613 (73)	—	30,913 (27)
1996	280,944 (100)	181,065 (64)	47,966 (17)	51,913 (18)
1997	335,030 (100)	222,310 (66)	63,225 (19)	49,495 (15)
1998	128,979 (100)	81,692 (63)	38,962 (30)	8,325 (6)
1999	241,176 (100)	144,319 (60)	64,924 (27)	31,933 (13)
2000 (Jan.–Oct.)	233,430 (100)	151,440 (65)	64,588 (28)	17,402 (7)

Sources: 1980–1990 extracted from MIDA (1998: appendix II) and Mahani (1997: 56); 1996–2000 extracted from MIDA (2001: 71)

Note
Number in parentheses indicates percentage of total

passenger vehicle imports, both in volume and in value terms. Imports of both CKDs and CBUs recovered in 1999, in both volume and value terms.

Component and parts manufacturers

Before the advent of the NCPs, the growth of the local component manufacturers was stifled by the lack of scale economies caused by the availability of numerous makes and models. Consequently, the component manufacturers became more interested in the replacement market. However, the introduction of the VDP shifted the component manufacturers' focus towards the needs of the local manufacturers and assemblers. According to MIDA (2001: 79), there are currently 350 automotive component manufacturers in the country, of which 52 per cent are Proton vendors while 9 per cent are Perodua vendors. The increase in the number of Proton vendors (the number of vendors multiplied by more than tenfold between 1985 and 1999) is shown in Table 4.9. The number of Bumiputera vendors has also grown very significantly, from four in 1985 to eighty-six in 1999. Component manufacturing in Malaysia has matured too, as approximately 70 per cent of the 350 companies involved are original equipment manufacturers (OEMs) for their global parent automotive producers (Singh 2000a: 5).

The growth in the number of suppliers has also resulted in an increased contribution to total employment in the automotive industry. Between 1985

Table 4.6 Passenger car sales in Malaysia and market share

Year	Type of cars				
	Total national cars	Proton	Perodua	Others	Total
1991	78,058 (64.2)	78,058 (64.2)	—	43,602 (35.8)	121,660 (100.0)
1994	119,385 (76.6)	110,505 (70.9)	8,880 (5.7)	36,380 (23.3)	155,765 (100.0)
1995	180,553 (80.2)	140,647 (62.5)	39,906 (17.7)	44,438 (19.8)	224,991 (100.0)
1996	223,041 (80.9)	176,100 (63.9)	46,941 (17.0)	52,574 (19.1)	275,615 (100.0)
1997	255,061 (82.8)	196,806 (63.9)	58,255 (18.9)	52,846 (17.2)	307,907 (100.0)
1998	126,410 (91.8)	87,489 (63.5)	38,921 (28.3)	11,281 (8.2)	137,691 (100.0)
1999	222,219 (92.7)	155,720 (65.0)	66,499 (27.7)	17,428 (7.3)	239,647 (100.0)
2000	261,444 (92.7)	178,960 (63.4)	82,484 (29.2)	20,659 (7.3)	282,103 (100.0)

Sources: 1991 extracted from Singh (2000a: 7); 1994–2000 extracted from Malaysian Motor Trades Association (17 July 2001)

Note
Numbers in parentheses indicate percentage of market share

Table 4.7 Exports of motor vehicles and parts from Malaysia

Year	Passenger vehicles		Motor vehicle parts
	Units	MYR (million)	MYR (million)
1992	18,849	330.6	65.7
1993	19,547	387.8	125.4
1994	14,806	313.4	174.4
1995	21,087	418.8	209.8
1996	21,748	475.4	211.7
1997	26,948	559.8	263.3
1998	23,700	667.8	314.3
1999	18,117	492.4	440.1
2000 (Jan.–Sept.)[a]	14,303	252.6	280.8

Sources: Singh (2000b: 8); MIDA (2001: 89)

Note
a Includes automotive and motorcycle parts

Table 4.8 Imports of motor vehicles and parts into Malaysia

Year	Passenger vehicles				Motor vehicle parts[a]
	CKD		CBU		
	Units	MYR (million)	Units	MYR (million)	MYR (million)
1993	139,271	1,338.3	12,648	192.8	503.7
1994	181,052	1,934.0	16,246	278.6	668.9
1995	262,566	3,043.4	17,812	385.4	256.1
1996	290,268	2,794.8	20,523	425.1	1,124.2
1997	363,181	2,795.3	11,608	289.0	1,416.8
1998	145,235	1,173.8	5,218	140.7	615.6
1999	263,696	2,244.4	5,470	146.8	1,100.2
2000	204,104	1,952.2	2,999	101.2	1,050.8

Sources: Singh (2000b: 8); MIDA (2001: 87–88)

Notes
a Includes automotive and motorcycle parts
CKD, completely knocked down; CBU, completely built up

Table 4.9 Proton vendor network, 1985–2000

Year	No. of vendors	Bumiputera vendors	No. of local parts[a]
1985	17	4	228
1986	33	7	325
1987	40	7	398
1988	46	9	525
1989	67	13	901
1990	78	21	1,014
1991	99	29	1,177
1992	106	35	1,316
1993	125	39	2,899
1994	128	42	3,444
1995	138	48	3,828
1996	151	71	4,076
1997	176	88	4,187
1998	188	93	4,319
1999	182	86	4,417
2000[b]	198	92	4,677

Source: Proton (2001: 10)

Notes
a Local parts = in-house + local + resourced.
b As of September

and 2000, employment in the motor vehicle parts and accessories sub-sector grew from 3,672 to 20,011. By 2000, its contribution to total manufacturing employment (0.66 per cent) exceeded that of the motor vehicles assembly sub-sector (0.57 per cent).

As shown in Tables 4.7 and 4.8, although exports of motor vehicle parts grew steadily from 1992 to 2000, imports increased even faster before the economy went into recession in 1998. The deficit in this sub-sector increased from MYR 356.1 million in 1992 to MYR 1,153.5 million in 1997. Subsequently, the crisis contributed to an improvement in the deficit to MYR 301.3 million as demand in the automotive sector contracted sharply with the fall in disposable income. After the crisis, the deficit grew back again to MYR 660.1 million in 1999, and, along with the recovery of the economy, grew even larger to MYR 770 million in 2000.

Impact on technology development of the automotive sector

Technology acquisition in the automotive sector is reflected in the number of technology agreements approved. In 1983, with the launch of the Proton, the number of technology agreements in the transport equipment sub-sector doubled from the previous year to twenty-two (Anuwar 1994: 114). By 2001, the technology agreements in this sub-sector numbered eighty-six, accounting for 13.4 per cent of the total number of agreements in the manufacturing sector. Only in the chemical sector (88) and the electronics sector (254) was the number of technology agreements higher than in the automotive sector (MIDA 2001).

However, while these agreements represent crude indicators of technology acquisition, they cannot in any way measure the technology transferred or absorbed by the recipients, since technology transfer involves more than the mere purchase of technology. While foreign technology is acquired when the technology-embodying factor (such as equipment, document, or technician) is imported, it is transferred only after it has been successfully adapted to local conditions, absorbed and integrated with the local economy. Therefore, technology transfer is a rather complex process and consequently there is no single indicator of its success. Rather, there are many ways to assess when it has happened. For example, success indicators of absorption include the ratio of actual rate of output to the designed rate, the cost of production, and various productivity measures. The assessment of technology transfer also needs to take into account the objective of technology transfer. Was the technology acquired to produce a product according to the design and specification of others, or to make minor design and technical changes, or was it acquired to reduce dependence on foreign technology? Since the development of local production capacity is one of the major objectives for the development of the automotive sector in Malaysia, this section will focus on the evolution of local content in Proton.

The first car produced by Proton was the Saga, which is essentially a modified version of the Mitsubishi Lancer. The Proton plant at that time carried out body stamping, assembly, painting and trim and final assembly of the car (Machado 1994: 301). Since bodies were added to imported, locally assembled kits, it was not surprising that the Saga had a local content of less than 20 per cent (Jomo 1994: 280). There was a strong preference instead to import components from Mitsubishi plants in Japan. This was due to the dependence on Mitsubishi for technology, which gave it a strategic negotiating position despite its minority share. As noted by Tham and Mahani (1999: 67), Mitsubishi's prerogative regarding the products to be made restricted Proton's ability to produce new models. At the same time, Mitsubishi's decisions on the types and quality of inputs used constrained Proton's control over the production costs, while the technology arrangements in linking vendors with technology partners also controlled the development programmes for parts suppliers.

Dissatisfaction with the pace of technology transfer led to the move to seek alternative technology partners. Thus while Mitsubishi technology was used to develop the Proton Saga, technology from Iswara, Wira, Perdana, Citroën of France and Rover of the United Kingdom was utilized for the development of the Proton Tiara and Satria (Abdulsomad 1999: 280).

Nevertheless, increasing pressures from the government, the Malaysian Automotive Component Parts Manufacturers' Association (MACPMA) and the high cost of importing from Japan due to the increasing value of the yen did facilitate improvements in local content over time. Table 4.9 shows that the number of local parts used in the production of Proton cars increased substantially from a mere 228 in 1985 to 4,076 in 1996. The reported local content achieved in the mid-1990s is about 54 per cent (Mahani 1997: 35). However, as cautioned by Leutert and Sudhoff (1999: 258), the actual local content may be lower than that stated figure, as basic input and raw materials are imported and vendors may use more than the 39 per cent imported input stipulated by the Generalized System of Privileges (GSP) regulations. More importantly, in terms of the technological sophistication of products, locally produced components were concentrated at the lower end, such as bodywork, accessories, wheels, tyres and electronic components. In contrast, engine parts, suspensions, shock absorbers and gear components were mostly imported.

The need to achieve greater and more coherent local production capabilities was certainly recognized by Proton. It was for this reason that Proton invested heavily in R&D. In 1992, Proton, as reported by Abdulsomad (1999: 295), spent about MYR 82 million on R&D. Subsequently, in 1993, a new Proton Research and Development Centre was established to carry out full-scale research on model making and computer-aided engineering design and manufacturing. At the same time, Proton's component and engine emissions testing laboratories were accredited by the UK Department of Transport to facilitate its export of cars to the United Kingdom. But since design, especially engine design, still depended on Mitsubishi, Proton made a strategic shift in 1996 by acquiring an 80 per cent equity stake in Lotus Group International Ltd. This move enabled Proton to obtain the vital design skills that it needed to design its own car.

Thus in 1996, Proton embarked on the first Proton-designed car project, the Waja. According to Proton (2000: 1), additional facilities costing almost

MYR 400 million were added to existing facilities to realize this project. The new hardware acquired included a rapid prototype centre, a computer-aided clay model milling machine, a prototype shop, a climatic chamber test laboratory, four dedicated engine test cells, a passenger safety sled test, four poster road simulators, roof crush and rollover testing, seat and seat belt anchorage tests for safety and a noise and exhaust emission laboratory. In addition, the component material strength and safety laboratory was enlarged. More than MYR 70 million was also invested to boost software areas such as computer-aided design, computer-aided engineering for design analysis on vehicle structure, computer-aided styling and a product management system. The number of research and design engineers were doubled from 150 to 300, while training was accelerated through on-the-job-training, as well as training from Mitsubishi and Lotus.

The design and engineering programme also required the supplier system to be restructured in line with global changes that favoured lean production. Under lean production, the relationship between the manufacturer and the supplier is preferably long-term and tight, in functional terms, and hence the design and production of components have to be executed with very close consultation between the two (UNCTAD 2000: 123). Consequently, more responsibility and risk is shifted to the supplier firms, with the most important trend being 'modular manufacturing', namely that suppliers manufacture and supply the entire 'sub-assemblies', or 'modules', rather than individual components only.

For the Waja, this shift was implemented with the identification of twenty local vendors as first-tier suppliers based on their design and development capabilities, quality, cost and reliability. These first-tier suppliers then designed their modules together with Proton engineers. Table 4.10 illustrates the increase in participation of the vendors in the design and development of the Waja as compared to the conventional system used in the development of the previous models. Unlike the conventional system, first-tier vendors have to assume responsibilities in prototyping and specification

Table 4.10 Responsibilities undertaken by first-tier suppliers

Activity	Modular/system	Conventional
Definition and specification	Proton	Proton
Designing	Proton/vendor	Proton
Development	Vendor	Proton/vendor
Prototyping and specification validation	Proton/vendor	Proton
Supplier sourcing (second tier)	Proton/vendor[a]	Proton
Manufacturing and process control	Vendor	Vendor
Testing and validation	Vendor	Vendor
Product approval	Proton/vendor	Proton
Mass production and production control	Vendor	Vendor
Warranty and after-sales services	Proton/vendor	Proton/vendor

Source: Proton interview conducted by the author on 16 July 2001

Note
a First-tier supplier will take 100 per cent responsibility after the Waja programme

validation, as well as supplier sourcing and product approval. As a result of this shift, the number of direct component suppliers used for the Waja model is only 108 vendors, compared with the 186 used for previous models. However, care was exercised not to displace the number of suppliers by this reduction, as Proton appointed them second-tier suppliers for the modular vendors.

Consequently, by the time the Waja was launched in 2000, the number of local parts increased to 4,677 (Table 4.9), while the local content in value terms supposedly increased to 80 per cent, as only the engine and transmission are imported from Mitsubishi. As in the case of the previous models, however, the modular suppliers also utilize imported materials, although they have to satisfy the GSP regulations. Hence actual local content should be lower than the touted 80 per cent. Moreover, each of the first-tier vendors is dependent on its foreign technology supplier and, out of the twenty, only twelve are Malaysian owned (Table 4.11). The other six vendors are multinationals in the global market, such as Robert Bosch (M) Sdn. Bhd.

Nevertheless, in the sixteen years since the first cars rolled off the assembly line at its plant in Shah Alam, Proton has come a long way. With the Waja, Proton has acquired intellectual property rights on its own developed platform, which is the under-body structure and chassis. This in turn can be used as a spin-off for a few other new products without incurring any new major costs. It has certainly, under heavy protection and various preferential treatments, fostered linkages that are among the most extensive in Malaysian manufacturing. However, as is discussed in the following section, there are equally, if not more, rapid changes in the external environment that may restrict Proton's development under protection in the future.

Current constraints

Three major factors that can affect the future development of the automotive sector in Malaysia are the government's commitments under the ASEAN Free Trade Area (AFTA), the World Trade Organization (WTO) and global trends.

ASEAN Free Trade Area

The coverage of products under AFTA is divided into three lists: the 'Inclusion List', the 'Temporary Exclusion List' and 'Unprocessed Agricultural Products'. Initially, automobiles and auto component parts were scheduled to be phased into the Inclusion List by 1 January 2000. While automotive component manufacturers adhered to this schedule, Malaysia instead deferred the phasing in for CKD and CBU vehicles to 1 January 2005 on the grounds that the crisis had a severely negative impact on the automotive sector. Ironically, as pointed out by Singh (2000b: 17), Malaysia's auto market suffered comparatively less damage than the Thai and Indonesian auto markets, and yet Thailand and Indonesia have chosen to liberalize their automotive markets.

Table 4.11 Waja GX first-tier (modular and system) supplier list

Module or system	Tier-1 supplier	Technology provider	Tier-2 supplier
Floor console module	Azman Hamzah Plastik Sdn. Bhd.	Survo, Japan	4
Front cross member module	Oriental Summit Industrie Sdn. Bhd.	Hirata, Japan	8
Fuel tank module	United Vehicle Industries Sdn. Bhd.	Kautex, Germany	3
Door module	Delloyd Industries Sdn. Bhd.	Kiekert, Germany	4
Instrument panel module	Hicom Teck See Manufacturing Sdn. Bhd.	Suryo, Japan	27
Pedal assembly	Tracoma Sdn. Bhd.	Mizushima, Japan	6
Seat assembly	Carseat (M)	Ikeda Bussan, Japan	17
Strut/absorber spring module	Sapura Automotive Industries Sdn. Bhd.		9
Brake system	TRW Automotive (M) Sdn. Bhd.	TRW Automotive, UK	5
Audio system	Clarion (M) Sdn. Bhd.	Clarion, Japan	3
Wiper system	Robert Bosch (M) Sdn. Bhd.	Bosch, Germany	2
Lighting system	Malaysian Automotive Lighting Sdn. Bhd.	ALMA, Germany	10
HVAC system	Denso (M) Sdn. Bhd.	Denso, Japan	12
	Patco Malaysia Bhd.	Calsonic, Japan	11
Alarm system	Multi-Code Electronics (M) Bhd.		2
Exhaust system	Automotive Industries Sdn. Bhd.	Sankei, Japan	11
Harness system	APM Industries Holding Bhd.		9
	Delphi Packard Electrics (M) Sdn. Bhd.	Delphi Packard, USA	9
Body sub-assembly	PHN Industries Sdn. Bhd.		
Safety restraints system	Autoliv Hirotaku Sdn. Bhd.	Autoliv, Sweden	

Source: Proton interview conducted by the author on 16 July 2001

World Trade Organization

Under the Trade-Related Investment Measures (TRIMS) Agreement, local content requirements were scheduled to be abolished by 2000. Malaysia has complied with this commitment for all sectors with the exception of the automotive sector, as the country has requested an extension of this deadline. It is uncertain, then, when the MDP and LMCP will be removed. Furthermore, subsidizing exports is also prohibited, so cars cannot be sold abroad at a lower price than those sold domestically after the deduction of transportation costs. This will have direct consequences for Proton exports, as domestic prices are reported to be higher than prices in the export markets (Rasiah 1996: 8; and 2001). More importantly, the automotive sector in Malaysia has not been subject to tariff binding or tariff reduction as yet (Pangestu and Stephenson 1996: 54), and any future negotiations in further reducing industrial tariffs will inevitably increase pressures to liberalize this sector.

Global trends

UNCTAD (2000: 125) distinguished five major trends that are affecting the future of the global automotive landscape. First, there is a trend towards increasing FDI in new facilities and modernizing old plants in countries with growth potential or lower costs of production. Second, transnational supply chains (TNSCs) are increasingly being reorganized to support lean production, leading to a reduction in the number of first-tier suppliers and an emergence of modular manufacturing systems. Third, TNSCs are moving towards the production of 'world cars' on global platforms that are compatible with a variety of car bodies. Fourth, design and R&D in parent companies are becoming increasingly centralized. Finally, the number of global players is actually shrinking with the recent wave of mergers and acquisitions.

Policy implications

Proton's installed capacity at 230,000 units per year has not changed since 1997, despite the economic recovery after the crisis in 1998 (Proton 2001: 6). The targeted production for 2001 is 210,000 units, hence it is operating close to full capacity. According to the *New Straits Times* (19 September 2001: B1), the multi-billion ringgit Proton City project that was deferred owing to the economic crisis in 1997 is scheduled to be revived and this will increase the production capacity of Proton, as the new plant in Proton City is projected to be able to produce up to 500,000 units annually by 2005 and a million cars by 2010.

However, the proposed expansion in domestic capacity will face heightened competition in both the domestic and the export markets because the delay in honouring its commitments under AFTA implies that much of the expected expansion in the AFTA market will be captured by other ASEAN members. Thailand in particular is poised to reap the gains from the expanded AFTA market as a result of its liberalization measures, and investments by global players that are positioning themselves in Thailand to take advantage of the AFTA market. In terms of liberalization, Thailand has moved far ahead of Malaysia, with tariff deregulation having been complemented by its adherence to removing LCRs in 2000. Thailand's firm commitment to liberalize the automotive sector was also accompanied by a warning that bilateral measures would be instituted if its neighbours did not comply with the scheduled deadlines (Singh 2000b: 14). Accordingly, Thailand insisted on compensation from Malaysia for the latter's postponing of the liberalization of its automotive sector. Thailand's plan to be the regional hub for the auto and auto parts industries is aided by both Japanese and non-Japanese investments. Consequently, Malaysia's delaying of liberalization to 2005 will mean that Proton will have to compete with the global players from Thailand, which will have made substantial inroads into the AFTA market by then.

At the same time, global players are also increasing their presence in Malaysia. For example, in July 2000, DRB-Oriental-Honda Sdn. Bhd (DOH) was formed with equity participation from DRB-HICOM Bhd. (36 per cent), Honda Motor Co. Ltd (49 per cent) and Oriental Holdings (15 per cent) to manufacture and distribute Honda cars in Malaysia (*The Star*, 13 June 2001, Business Section: 3). The manufacturing plant in Malaysia was scheduled to roll out the first Honda car unit in early 2003. Since this is the fourth Honda plant in ASEAN – after those in Thailand, Indonesia and the Philippines – Honda is also positioning itself for the AFTA market.

Therefore, with liberalization, Proton's domestic market share will undoubtedly shrink, while it is unclear whether it can successfully compete for the AFTA market. There is clearly a need for Proton to enhance its competitiveness in view of the impending liberalized environment. Two key issues that will determine its ability to compete are its production costs and the extent of technology upgrading required in order for Proton to transform itself from a domestic manufacturer into an internationally competitive firm.

In terms of cost, the previous practice of single-sourcing and protectionism has constrained the cost competitiveness of Proton to such an extent that Proton's production costs are estimated to be 20 per cent higher than those of its international competitors (Mahani 1997: 41). While multi-sourcing will increase competitiveness and eventually lower costs for Proton, the current volume of production remains a severe constraint for cost efficiency, unless parts and component manufacturers increase their supply for other makes within the country and, more importantly, move towards the export market. It appears that current policy directions are encouraging this shift. First, liberalization of the automotive component branch will precede the liberalization of the CKDs and CBUs, as this segment is adhering to the AFTA schedule and the local content policy will eventually be phased out. Second, the new vendor programme emphasizes links with international automotive component manufacturers, R&D, design engineering activity and design ownership (MIDA 2001: 98). The stated development strategy also includes promoting the relocation of component manufacturers, especially international first-tier suppliers and System Integrators (MIDA 2001: 95). Hence domestic vendors will have to form partnerships with international first-tier suppliers to penetrate the export market, or be second-tier suppliers to the international first-tier suppliers that will enter the Malaysian market when protection has been removed.

Volume can also be increased if Proton exports more than its current amount. For Proton, exports peaked in 1997, when they constituted 13 per cent of total production and Proton exported mainly to the UK and Irish markets (Proton 2001: 8). Continuing to increase exports, especially without the price differential between domestic and export prices, will pitch Proton against the established global automotive giants in established markets and emerging manufacturers from Korea and Brazil for the less established markets. It will also require considerable technology upgrading. Despite the contribution made by Lotus towards enhancing the technology development of Proton, this may still be insufficient. As noted by Rasiah (2001: 17), Lotus technology differs substantially from Proton's range of cars, as Lotus caters more for the luxury niche, while Proton targets the lower- and middle-class consumer. Hence Proton may not be able to adapt the Lotus technology in time when the market is scheduled for liberalization in 2005. For example, the Malaysian-designed Waja continues to utilize a Mitsubishi engine, while the Malaysian-designed engine is still a work-in-progress, scheduled to come on-stream in 2003. Moreover, it is not known whether the Malaysian-designed engine can satisfy the requisite safety and environmental standards for export markets. Therefore, Proton will still need to form a strategic alliance with one of the leading auto companies in order to compete with the entry of more global players into the impending liberalized Malaysian market and to enter the AFTA and world markets.

Thus dismantling protection will require policy measures that can ease the cost and volume constraints that are affecting both the vendor system and Proton. These constraints, together with that on technology, imply the need to form strategic partnerships for domestic vendors as well as for Proton. Proton's own strategy to overcome both cost and volume constraints (*The Star*, 9 January 2002, Business Section: 3) involves producing auto parts in China so as to lower the cost of production and secure for itself a strategic position for meeting the projected increase in the demand for cars in China.

Conclusion

The automotive sector in Malaysia represents an interesting case study for analysing the contributions of state intervention to industrial development. The sector has certainly received massive support from the government, support that has gone beyond the use of tariffs in the classic infant industry argument. Its policies have enabled Proton to become one of the biggest car manufacturers in ASEAN and have nurtured a large local parts industry, thereby fulfilling, to some extent, the objectives of the NCPs.

However, local design and development capabilities did not take off until the acquisition of Lotus in the second half of the 1990s and the production of the Malaysian-designed car, the Waja. Even though this has accelerated the learning curve of Proton, it was implicitly insufficient, as the government has delayed the liberalization as applied to CBUs and CKDs under AFTA. It is also unlikely that the delay to 2005 can hasten Proton's technological development to such an extent that a completely Malaysian-designed car (including engine and transmission) can compete at internationally competitive prices, since volume constraints, in particular, pose a formidable barrier unless Proton can successfully increase its exports.

Further delay in liberalization will not improve the situation, as most of the AFTA market will be captured by other ASEAN car manufacturers, especially the global players that are already operating in Thailand. Loss of the ASEAN market will entail moving outwards to the global market, which will be more difficult to penetrate as safety and environmental standards can act as even more severe barriers to trade than tariffs. Liberalization will, however, ultimately imply the need to forge strategic alliances such as mergers with existing global players.

The development of the automotive sector in Malaysia thus raises several pertinent lessons regarding the role of state intervention. First, there should have been a greater focus on technology transfer from the very beginning. The slow transfer of technology was attributed, in part, to the lack of detailed agreements on the pace and mechanisms to monitor the transfer of technology in the initial contract between HICOM and the technology partner, Mitsubishi (UNCTAD 2000: 165). Second, there should have been a scheduled phase-out of protection. Since protection in the infant industry argument is meant to be temporary and to help the infant to 'grow', it should not have been left to the forces of liberalization and globalization to enforce the opening up of this sector. Third, since technological development is crucial for the survival of this industry in the global market, protection should have been accompanied by a suitable focus on human resource development that would provide the engineers and designers that are required for developing indigenous design and development capabilities.

References and further reading

Abdulsomad, K. (1999) 'Promoting Industrial and Technological Development under Contrasting Industrial Policies', in Jomo, K.S., Felker, G. and Rasiah, R. (eds) *Industrial Technology Development in Malaysia*, London: Routledge.

Alavi, R. (1996) *Industrialisation in Malaysia: Import Substitution and Infant Industry Performance*, London: Routledge.

Anuwar, A. (1992) *Malaysia's Industrialization: The Quest for Technology*, Singapore: Oxford University Press.

—— (1994) 'Japanese Industrial Investments and Technology Transfer in Malaysia', in Jomo, K.S. (ed.) *Japan and Malaysian Development: In the Shadow of the Rising Sun*, London: Routledge.

Chee, P.L. (1992) 'Managing Structural Change and Industrialisation: Heavy Industrialisation – The Malaysian Experience', paper presented at the ISIS–HIID Conference on the Malaysian Economy, ISIS, Kuala Lumpur.

Jomo, K.S. (1994) 'The Proton Saga: Malaysian Car, Mitsubishi Gain', in Jomo, K.S. (ed.) *Japan and Malaysian Development: In the Shadow of the Rising Sun*, London: Routledge.

Leutert, H.G. and Sudhoff, S. (1999) 'Technology Capacity Building in the Malaysian

Automotive Industry', in Jomo, K.S., Felker, G. and Rasiah, R. (eds) *Industrial Technology Development in Malaysia*, London: Routledge.

Machado, K.G. (1994) 'Proton and Malaysia's Motor Vehicle Industry', in Jomo, K.S. (ed.) *Japan and Malaysian Development: In the Shadow of the Rising Sun*, London: Routledge.

Mahani, Z.A. (1997) 'The Automotive Industry in Malaysia', Final Report submitted to the Asian Development Bank Study on the Manufacturing Sector Competitiveness in ASEAN.

Malaysia (1996) *The Second Industrial Master Plan: 1996–2005*, Kuala Lumpur: Ministry of International Trade and Industry.

Pangestu, M. and Stephenson, S. (1996) 'Evaluation of Uruguay Commitments by APEC Members', in Bora, B. and Pangestu, M. (eds) *Priority Issues in Trade and Investment Liberalisation: Implications for the Asia Pacific Region*, Singapore: Pacific Economic Cooperation Council.

Proton (2001) *Corporate Information Booklet*, Shah Alam: Proton.

Rasiah, R. (1996) 'Rent Management in Malaysia's Proton', *IKMAS Working Papers*, no. 2, Bangi, Malaysia.

—— (2001) 'Liberalisation and Car Manufacturing in SEA-4', *Business and Society*, 2 (1): 1–23.

Singh, P. (2000a) 'The Malaysian Automotive Industry', *The ASEAN Automotive Industry: Challenges and Opportunities*, AUSPECC. Online at http://www.asean-auto.org/mal/report.htm (accessed 9 January 2003).

Singh, P. (2000b) 'Impact of Regional Trading and Industrial Arrangements in the ASEAN Automotive Industry', paper presented at the Regional Conference on the ASEAN Automotive Industry, Manila.

Tham, S.Y. and Mahani, Z.A. (1999) 'Industrial Institutions: The Case of Malaysia', in Barlow, C. (ed.) *Institutions and Economic Change in Southeast Asia: The Context of Development from the 1960s to the 1990s*, Cheltenham, UK: Edward Elgar.

UNCTAD (2000) *The Competitiveness Challenge: Transnational Corporations and Industrial Restructuring in Developing Countries*, Geneva: UNCTAD.

Unpublished internal company documents

MIDA (1998) 'The Automotive Industry in Malaysia', Kuala Lumpur: Transport and Machinery Industries Division.

—— (2001) *Industry Brief* (February), Kuala Lumpur.

Proton (2000) 'News Release' (8 May).

Personal communications

(2001) Interview with Proton conducted by the author (16 July).

(2001) Interview with MIDA conducted by the author (24 July).

5 Philippine policies for the automobile sector

Focus on technology transfer

Gwendolyn R. Tecson

Like most developing countries, the Philippines had been enthralled by the possibility of developing its own automobile industry. Not only was this considered a 'prestige industry', it was also thought to possess the characteristics of an industry in which, with the strong backward and forward linkages it could create, possibilities for technological learning were large. It could help spur development in many upstream industries, such as in materials, metals, machinery, chemicals and the electrical and electronics industries. Technological learning was also made possible, so it seemed, by the industry's standardized level of technology. This made it less costly for investors from developed countries to transfer technology and for recipients from developing countries to absorb it. Thus the car industry easily became a target for industry development via import-substitution protection policies in many developing countries, the Philippines included.[1]

The country's development objectives found an echo in the needs of multinational automotive firms, which were faced with increasing global competition to invest overseas. They were in constant search for new markets, low-cost assembly points, as well as component suppliers that would enhance their competitive advantage in export markets. In particular, Japanese firms that had fine-tuned their subcontracting networks in Japan were able to take advantage of ASEAN developing countries' desire to jump-start their local automotive industries. For instance, Toyota's vehicle production network in the Asian region, while created partly as a result of each of these countries' import-substituting policies, was a rational response to the regional industrial cooperation policies of ASEAN. Thus Toyota's intra-firm trade in parts and components throughout the region, coordinated by Toyota Motor Management Company, located in Singapore, leverages the comparative advantage of each country in the region. Toyota exports steering links from Malaysia, diesel engines from Thailand, transmissions from the Philippines and engines from Indonesia (UNCTAD 1996: 11).

What policies have been instituted by the Philippine government, as host country, to multinational automotive firms in order to encourage the transfer of technology? How and to what extent have these multinational enterprises (MNEs), especially the Japanese transplants in Asia, been induced by policy to contribute to the technological learning of the local firms? In what follows, we shall inquire into the role of public policy in inducing the transfer of technology to host firms and the response of MNEs to the said policy.

The outline of the study is as follows. The first section examines the evolution of government policy for the automobile industry in the Philippines, particularly in terms of the expectations it harboured for technology transfer by MNEs, as well as the extent to which its goals were realized. The second section examines the reasons for the limited effects of policy on the development of the automobile industry in the Philippines, both in assembly and in component manufacturing. The third section discusses the extent of technology transfer taking place in Japanese firms operating in the Philippine automotive industry, on the basis of the results of discussions with a number of managers in the industry. Two modes of technology transfer have been considered, namely international transfer (through the development of local industry via foreign direct investment (FDI) and domestic diffusion. The final section concludes the study with an evaluation of the significance of government policy on technology transfer in the Philippines and attempts to discover the most important lessons from such an experience on industrial development of the country.

Philippine automotive industry development policy and technology transfer

The history of the automotive industry in the Philippines is a classic example of the use of import substitution policy for industry development. There was no local car industry to speak of during the immediate post-Second World War period, but the subsequent balance of payments crisis in 1949 gave the government a chance to take advantage of the nascent domestic demand for vehicles to develop a local automobile industry. Thus it prohibited the importation on a commercial scale of completely built up (CBU) cars and granted foreign exchange allocation only for the importation of completely knocked down (CKD) components. This move was intended to encourage the setting up of assembly plants locally. Ten years later in 1960, there were twelve local vehicle plants turning out thirty different brands from Western Europe, the United States, Australia and Japan. Eight years hence, in 1968, the number of plants had risen to twenty, assembling sixty different kinds of models. With local annual demand of only about 17,000 units, no plant could be anywhere near the minimum efficient scale.

The government, forced to 'rationalize' the industry, instituted a Progressive Car Manufacturing Program (PCMP), which limited the number of assemblers to only five.[2] The policy carrot was the continued prohibition of CBU imports, which allowed cars assembled domestically to sell at prices much higher than international prices. But the stick was a local content policy, which required assemblers to source a portion of their parts and components domestically. A 10 per cent domestic content in 1973 was expected – quite unrealistically – to rise to 62.5 per cent (eventually scaled back to 52.5 per cent) three years later. Through this local content policy, multinational assemblers, which had by then teamed up with locals (in view of the constitutional restrictions on foreign equity participation of forty–sixty in favour of Filipinos), were to spur the domestic manufacture of automotive components. The Board of Investments (BOI) expected them to subcontract the manufacture of parts and components to local producers, particularly to small and medium-sized enterprises (SMEs), and in the process 'upgrade engineering and production skills and provide new technological know-how to the country's industrial sector', according to an internal document of 1973. It was hoped that Japanese assemblers, especially, would transplant the practice of subcontracting prevalent in Japan, a practice considered to have been the source of Japanese competitive strength in automotive production. For instance, in Japan, Toyota had adopted the practice of contract assembly since the late 1940s (Nishiguchi 1994: 104).

It was believed that Toyota transferred technology to its subcontractors[3] in the process of subcontracting components sub-assembly. Something similar was expected to take place in the Philippines as Japanese assemblers began to invest in the country.

Policy objectives versus realizations

As provided for by the PCMP, beginning in 1973 five assembly firms were admitted to operate and were allowed to import CKD units. In order to meet the local content targets, the assemblers, under prodding by the BOI, began to manufacture parts: for example, engine blocks, coil springs and springs (Delta Motors); body stamping and soft trims (Ford Philippines); transmissions (Yutivo Francisco (GM-Isuzu)); transmissions, soft trims and wiring harnesses (Chrysler Philippines-Mitsubishi); and seat pads and miscellaneous car parts (DMG Inc.). The second requirement of the PCMP stipulated that firms should export so that the participants could earn part of their foreign exchange requirements by importing CKDs. To comply, the participants then began to export their manufactured parts (Abrenica 2000a: 88): engines to Japan (Delta Motors); body panels to other ASEAN countries (Ford Philippines); transmissions to Australia and ASEAN countries (Yutivo-Francisco); transmissions to Japan and ASEAN countries (Chrysler Philippines); and seat pads and other car parts to Germany (DMG Inc.).

However, instead of subcontracting these parts to local Filipino firms, particularly SMEs, as envisioned by the policy makers, the five participants opted instead for in-house production of parts or, in some cases, for the setting up of subsidiaries partly owned either by the assembler or by a supplier of the assembler. An example of the latter is Hella, a subsidiary of Hella Inc., a German multinational specializing in auto lights and horns. In the case of the Japanese assembler Toyota, which was used to relying on a network of suppliers back in Toyota City, Japan, local content was provided in-house or by affiliate firms that came to invest in the Philippines, as for example Aichi Forging, a member of the Toyota *keiretsu*, did in 1974. Thus things did not quite progress the way policy makers envisioned they would. Moreover, by 1978, five years after the launch of the PCMP, local content was still below 30 per cent.

While the production of higher value added components was kept either in house or within a tightly controlled affiliate, some lower value added parts began to be produced by local (Filipino) firms. There were more than 220 local (Filipino) parts suppliers in 1978, as compared with only thirty-two at the start of the PCMP in 1972.[4] A government evaluation study of the PCMP reported that technology transfer was realized in several manufacturing processes, such as machining, gear cutting, forging, die casting (aluminium) and heat treatment. However, a 1996 study by the Department of Science and Technology (DOST) revealed that the 1995 level of ductile iron technology in the Philippines had not yet even reached Japan's level from 1965. And while Thai foundries are said to produce a minimum of 120 moulds per hour per machine, the most progressive foundries in the Philippines were found to produce only a maximum of 20–25 moulds per machine (cited by Abrenica 2000a: 85). The country's metalworking industry is said to be quite unable to deliver the outputs required by assemblers. These latter, then, have been forced to integrate vertically and to undertake other strategic alliances even with their competitors. For instance, Toyota Motors Philippines Inc. is said to have produced forged parts for transmissions in a joint venture with Philippine Automotive Manufacturing Corporation (PAMCOR, assembler of Mitsubishi vehicles). Moreover, a study that evaluated the PCMP concluded that technology transfer was not substantial. Local technicians were being trained to operate imported machinery, so technology transfer was limited to a transfer of machinery and equipment, with training and continued checkups being undertaken by foreign consultants. Thus assimilation or absorption of technology was not encouraged at the time.

An important factor cited by the BOI that prevented the achievement of the desired percentage of local content was the use of 'deletion allowance'. This allowed foreign firms to transfer prices and, in the process, discourage local sourcing. When a part or component was used in an assembled unit, it was assigned a 'deletion allowance', i.e. the value to be deducted from that of the CKD pack. Since this was based on the marginal production cost of the parent company (excluding profits), it was priced lower than if the component were to be imported separately – that is, when not included as part of the CKD pack. Thus the more parts sourced locally, the higher the price of the locally assembled vehicle (Ken 1977, cited by Abrenica 2000a: 89–90), making CKD importation a more attractive alternative to local sourcing. As will be discussed later, the assemblers were wont to use other options provided as 'loopholes' in the local content policy.

The deep economic crisis in 1983, however, plunged the entire automotive industry into bankruptcy (Delta Motors) and eventual flight from the Philippines (Ford, GM, Chrysler and DMG). Even two Japanese firms that stayed behind, Pilipinas Nissan Inc. (which had taken over DMG) and PAMCOR, closed down in 1986 when parts importation was restricted as a result of the severe foreign exchange crisis. By then, the number of parts manufacturers had shrunk to only about forty.

Renewed attempts at car development

After the smoke cleared somewhat in 1987 with the assumption of power by the new Aquino administration, the PCMP was resurrected under the Motor Vehicle Development Program (MVDP). The latter programme was supposed to be different from the previous one in terms of its greater emphasis on the development of automotive parts manufacturing. It retained the ban on importation of CBU units as well as the local content programme.[5] Assemblers were obliged to invest in parts manufacturing within three years, amounting cumulatively to at least 9 per cent of the net local content requirement under the MVDP. Likewise, they were to source their foreign exchange requirements increasingly through exports of automotive parts.[6]

Recognizing the problem of developing a viable parts industry in the face of market fragmentation, the government, under the Car Development Program (CDP), initially limited both the number of assemblers – only three were allowed to participate, namely Nissan, Toyota and Mitsubishi – and the number of their models, with each assembler allowed up to three basic models and two variants per basic model. Furthermore, joint venture projects were recognized as an important vehicle of technology transfer. Thus the BOI was allowed to grant a foreign exchange concession to new joint venture car assembly operations, which would be set up under the amended CDP amounting to 10 per cent of the net foreign exchange earnings (NFEE) required during the first two years of operation of such new joint venture companies, provided that they had at least 30 per cent local equity. Moreover, additional foreign concessions amounting to 10 per cent of the NFEE of parts manufacturing could be granted to new or existing participants who would promote joint ventures between their parts vendors and local autopart manufacturers. Over a ten-year period, i.e. by 1998, it was expected that the local car parts industry would achieve a competitive status, and that local parts producers would absorb technology successfully.

Deregulation and proliferation of makers

However, the dust had hardly settled on the new policy approach when changes were introduced in close succession, each with a potential impact on the possibilities for technology transfer.

- In January 1990, a People's Car category was opened to which seven assemblers were admitted (each allowed one basic model with two variants); after a one-year assembly operation, these were allowed to enter the main category,[7] leading to the entry of five of the original seven participants and, eventually, to the 'proliferation of more brands and makes of vehicles than there were during the PCMP period' (Abrenica 2000a: 93). This meant an even greater fragmentation of the domestic market, making parts production less standardized and hence more costly to undertake.
- In December 1992, a luxury car category was opened under the CDP, giving access to European assemblers (Volvo, Mercedes-Benz). While

these were not subjected to a local content requirement, they were expected to invest US$8 million in parts and component manufacturing and to generate 100 per cent of their foreign exchange requirements from exports of parts.

- In May 1993, there was a relaxation of the import restrictions on brand new luxury cars,[8] while the number of basic models for each assembler in the main car category was increased from three to four.

- In 1994, the ASEAN Industrial Joint Venture project was accommodated, leading to the entry of Malaysia's Proton (Perusahaan Otomobil Nasional Sdn. Bhd., a joint venture with the Autocorp Group).

- In 1995, the ongoing manufacturing-wide trade reform policy led to a widening of the tariff differential between assembly and CKD importation from 20 to 37 per cent. This was due to the simultaneous increase in tariffs on CBUs from 30 to 40 per cent, given the tariffication of the former quantitative restrictions on CBUs under the World Trade Organization (WTO) and the simultaneous decrease in tariffs on CKDs from 10 to 3 per cent. However, tariffs on raw material imports to locally produced parts ranged from 10 to 35 per cent. With this policy, it became clear that car assembly was being favoured over car parts manufacturing, notwithstanding the stated programme objective of developing a competitive local car parts industry. The widening tariff differential accorded car assembly greater effective protection. Local parts industry development was compromised in favour of importing CKD packs, and possibly the manufacture of certain parts was penalized with reduced, if not negative, effective protection. It was hoped that the bias against parts manufacturing would be corrected by the forging of an agreement two months later that encouraged assemblers *voluntarily* to source at least 40 per cent of parts requirements locally. But since it was voluntary, no definite commitments came of it. The differential is to disappear only with the uniform tariff of 5 per cent on all CBUs and parts imports in 2004, marking the end of the tariff reform programme.

- In March 1996, all import restrictions on brand new CBU cars, motorcycles and light commercial vehicles (LCVs) with a seating capacity of fewer than ten passengers were lifted, and so were the limits on the number of models and variants of models for car assemblers, as well as the five-year period of maintaining the models. Moreover, price ceilings on all automotive vehicles, with the exception of the People's Car, were removed. The government also did away with the rule that required assemblers to adjust their local content progressively. However, they were still required to maintain a local content ratio for passenger cars of 40 per cent and for commercial vehicles depending on the category, from 13.8 to 54.8 per cent. A new assembler could be accredited so long as it met the investment requirements. Even the mandatory deletion list was removed (Abrenica 2000a: 94).

Then, in 1998, changes were introduced again into the programme guidelines, the two most important being the allowing of foreign exchange earnings from exports of automotive parts worth at least US$200 million annually to be considered for local content credit, and restoration of the mandatory list of components to be locally sourced. In the case of the second amendment, the BOI was vested with the power to adopt a short mandatory list of strategic and high-technology automotive parts (e.g. engines, transmissions, clutch and brake systems, suspension systems and drive-line assemblies). Assemblers would then be prevented from importing these parts after the set target dates for local production. The expectation of this policy was that it would promote technology transfer in the long run by inducing foreign carmakers to purchase their major automotive parts locally instead of importing them, thus encouraging FDI in the production of these high-technology parts and, consequently, stimulating the transfer of technology. The restoration of the mandatory list was thus premised on the expected role FDI would play in upgrading the local industry's technological levels and generating higher value added. By 2000, however, both the local content requirement and the foreign exchange requirement were terminated following the country's commitment to the Agreement on Trade-Related Aspects of Investment Measures (TRIMs Agreement) under GATT–WTO requirements.

FDI and exports in the automotive industry

Recent pronouncements by the Philippine government, then, seem to indicate policy changes towards laying greater emphasis on attracting more investments from global players into assembly and car parts manufacturing. Increased FDI enhances the possibilities for transfer of technology from MNEs to local manufacturers, in addition to other 'software' that FDI brings with it (such as management, marketing, financing and knowledge of export markets). Table 5.1 shows the inflows of FDI into the automotive industry in recent years.

However, compared with the inflow of FDI into manufacturing, the automotive industry seems to be receiving less over time. Given the difficulty of

Table 5.1 Inward foreign direct investments stock in the Philippine automotive industry, 1973–2000

Year	Transport equipment	Manufacturing	Share (%)
1973–1989	109.01	1,484.38	7.34
1990	8.18	129.10	6.34
1991	45.59	414.87	10.99
1992	40.00	206.30	19.39
1993	13.74	309.75	4.44
1994	5.66	680.39	0.83
1995	53.03	337.87	15.70
1996	35.70	477.69	7.47
1997	17.47	172.19	10.15
1998	6.53	245.55	2.66
1999	21.44	1,049.16	2.04
2000	6.06	171.67	3.53
1990–2000	253.41	4,194.53	6.04

Source: Bangko Sentral ng Pilipinas: *Selected Philippine Economic Indicators*, various issues

finding a viable supplier network among local firms, a major approach to complying with the local content rule by Japanese assemblers was to bring in Japanese supplier firms. Table 5.2 shows the dates of registration of Japanese assembly firms and their suppliers in the Philippines. According to these data, Japanese supplier firms were established mainly in the 1990s. Why only in the 1990s in spite of the application of local content policy since the start of the PCMP in the 1970s? There are a number of reasons for this. Although Japanese firms have been investing overseas since the 1950s, the large FDI outflows took place mainly after the Plaza Agreement of 1985. However, unlike other ASEAN countries that were politically and economically ready at the time to serve as hosts to Japanese FDI, the Philippines was not. It was only sometime after 1993 that the Philippines had been able to recover from the devastating political instabilities that caused a series of *coup d'état* attempts that plagued the post-Marcos Aquino government. Also, it took a sharp yen revaluation to force out Japanese firms – particularly the small and medium-sized component manufacturers – in the automotive industry, which had so far been able to maintain comparative advantage while exporting from Japan.

Another important reason for the increased inflow of FDI is the formation of the regional free trade area in ASEAN, AFTA, which allowed Japanese firms to take advantage of declining tariff rates on automotives in ASEAN countries and to strengthen their network of supplier firms. Even before the formation of AFTA, the brand-to-brand complementation scheme had benefited the trade of auto parts by the Japanese assemblers in the different countries of the region, which took advantage of preferential import rates. More recently, components trade within ASEAN is accelerating under the AFTA preferential rate of 0 to 5 per cent with 40 per cent ASEAN content. Maximum component tariffs are expected to be 20 per cent for AFTA trade in 2000 and will decline to a maximum of 5 per cent by 2003. Within AFTA, immediate access to the preferential tariff rates of 0 to 5 per cent is possible under AICO (ASEAN Industrial Cooperation), which the Philippines has approved for various manufacturers in the auto sector. Figure 5.1 shows Toyota Motors' brand-to-brand and AICO schemes.

Aside from the emphasis on attracting FDI into the industry, the target of the medium-term development plan is for car parts exports to generate US$2 billion worth of export revenue by 2004, up from only US$323 million in 1998. Table 5.3 shows that, although the country remains overall a net importer of automotives, net exports in parts and components were recorded in 1998, as exports began to exceed imports in that year (Figure 5.2).[9] A breakdown of exports in 1998 shows that the main exports consist of servo breaks, gearboxes (destined mainly for Germany), transmissions and wiring harnesses. It is possible that this is a short-term response to the foreign exchange and local content requirement. It must be recognized, however, that exports constitute an effective way out of the problem of high cost associated with a limited and fragmented domestic market, because production for export markets can provide the necessary economies of scale. The Taiwanese model of a very limited local assembly but a well-developed export-oriented parts and components industry is fast becoming a viable

Table 5.2 Date of registration of automotive assemblers and component manufacturers in the Philippines, 2000

Major assembler	1960s and 1970s	1980s	1990–1995	1995–2000
Honda	Honda Philippines, Inc. (73)	Tsuchiya Kogyo (89.8)	Honda Cars Philippines, Inc. (90.1) Ohtsuka Polytech (90.0) Honda Engine Mfg Phil. (92) Honda Parts Mfg Corp. (92) FCC (Philippines) Corp. (93.9) F. Tech Phil. Mfg Inc. (94.5) Morikoku Philippines (94.5) Kosei Asia Pacific (94.8) TS Tech Trim Phil. (94.9) Kawashima Textile (94)	Yutaka Mfg Phil. (95.4) Nissin Brake Phil. Corp. (95.5) Hadsys Phil. Corp. (95.7) Laguna Metts Corp. (95.12) TS Tech Philippines (96.3) Imasen Phil Mfg Corp. (96.7) Bestex Okaya Philippines (96)
Toyota	Aichi Forging (74.5) Phil. Koyo Bearing Corp. (75.7)	Toyota Motor Phil. Corp. (88.8) Nachi Pilipinas Ind. (89.1)	Toyota Autoparts Phil. (90.8) Fujitsu Ten Corp. (90.8)	Phil. Auto Components (95.3) TRP Inc. (95.7) Takanichi Phil. Corp. (96.3) Jeco Autoparts Phil. Inc. (96.9) Techno Eight (98)
Mitsubishi	Asian Transmission Corp. (73.1)	Mitsubishi Motors Phil. Corp. (87.2)		PNC (96)
Nissan	Sanoh, Fulton Inc. (79.12)	Nissan Motor Phil. (83.6) Clarion Mfg Corp. (89.9)	Orion Rubber Mfg Corp. (91) Clarion Mitsuwa Phil. Inc. (92.3)	Nissan Diesel Phil. Corp. (95.10) Fas Cebu Corp. (96.1) Tennex Phil Corp. (96.5) Valtech Rubber Indl Corp. (98)
Isuzu	Kawasaki Motors (68.6) Pilipinas Hino (75.9)	Suzuki Philippines (85.2)		Isuzu Philippines (95.8) Isuzu Autoparts Mfg Corp. (96.11)

Source: Interview Survey, Institute of Development Economies, 1999, and from various sources

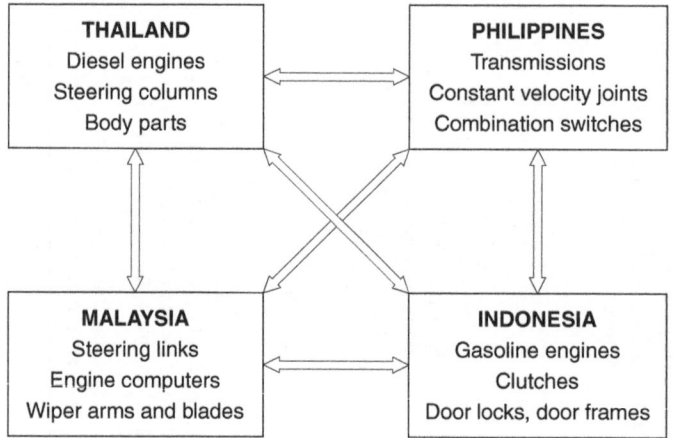

Figure 5.1 Toyota Motors' brand-to-brand complementation and the ASEAN Industrial Cooperation (AICO) network

Source: Toyota Motors, Co. (Findlay and Abrenica 2000: 3, Figure 5)

model to follow. Table 5.4 shows that the strategy of Japanese MNE assembly firms for achieving scale economies is that of exporting specific automotive parts and components from their different subsidiaries and affiliate companies in ASEAN. Toyota, Mitsubishi and Isuzu Motors are showing a preference for exporting transmissions from the Philippines to the different ASEAN countries. Also evident is the extensive development of the supply network of Japanese automotive investors in Thailand. Such a specialization in part and component manufacturing by the Japanese MNEs leverages the comparative advantage of different countries in the region, taking into consideration not only labour cost and raw material availability, but also cost advantages from declining learning curves, past investments in training workers and availability/cost of technical labour and engineers, etc. A similar picture showing Toyota's two-way trade in parts and components within ASEAN (including Taiwan) is shown in Figure 5.3.

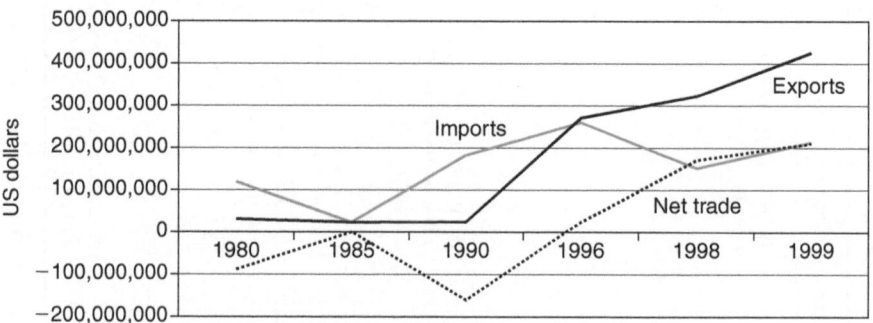

Figure 5.2 Trade in automotive parts and accessories in the Philippines, 1980–1999

Source: National Statistical Office: *Foreign Trade Statistics of the Philippines,* various issues

Table 5.3 Trade in automotives in the Philippines

Item	1980	1985	1990	1996	1998	2000
Road vehicles						
Exports	1.28	0.04	1.10	3.84	11.81	484.80
Imports	80.06	18.48	278.17	1,059.59	226.30	796.82
Net trade in road vehicles (US$m)	(78.78)	(18.44)	(277.07)	(1,055.75)	(214.49)	(312.02)
Parts and accessories						
Exports	30.25	20.98	20.85	269.64	323.78	425.18
Imports	118.33	22.76	179.53	261.64	153.14	212.53
Net trade in parts and accessories (US$m)	(88.08)	(1.76)	(158.68)	8.00	170.64	212.65

Source: *Foreign Trade Statistics of the Philippines*, NSO: various issues

Table 5.4 Regional component flows in ASEAN by major Japanese assemblers

Origin \ Destination	Thailand	Indonesia	Malaysia	Philippines
Thailand		**Toyota:** engine, pressed parts, alternators, starter, steering coil, in-panel, stabilizers **Mitsubishi:** — **Honda:** side panel, flow panel, door panel, trunk hood, right handle for City **Nissan:** high pressure cable, pressed parts, interior trim, rear combination, water pump, oil pump, radiator **Isuzu:** diesel engine, pressed parts, engine parts	← engine parts (↑, as Indonesia)	**Toyota:** engine, pressed parts, alternators, starter, steering coil, in-panel, stabilizers **Mitsubishi:** engine, L200 parts
Indonesia	**Toyota:** as in Mal. & Phil. **Mitsubishi:** brake **Honda:** as in Mal. & Phil. **Nissan:** meter **Isuzu:** AUV parts, brake parts		**Toyota:** 7 K engine, clutch, door frame, seat adjuster **Mitsubishi:** — **Honda:** cylinder block, cylinder head **Nissan:** — **Isuzu:** —	**Toyota:** 7 K engine, clutch, door frame, seat adjuster **Mitsubishi:** body **Honda:** cylinder block, cylinder head **Nissan:** — **Isuzu:** —
Malaysia	**Toyota:** steering link, engine computer, aircon, joint, pressure relay, antenna **Mitsubishi:** steering coil **Honda:** bumper, in-panel **Nissan:** pressed parts, spring suspension **Isuzu:** steering gear	**Mitsubishi:** steering coil (↑) **Honda:** bumper, in-panel (↑) **Nissan:** — **Isuzu:** —		**Toyota:** as in Thai. & Mal. **Mitsubishi:** as in Thai. & Indonesia **Honda:** as in Thai. & Ind. **Nissan:** as in Thai. **Isuzu:** —
Philippines	**Toyota:** transmission, joint, combination switch, steel parts **Mitsubishi:** transmission **Honda:** intake manifold, coil, console, pedal, converter, right handle for City **Nissan:** pressed part, ventilator, pedal waist **Isuzu:** manual, transmission	**Toyota:** transmission → (↑) **Mitsubishi:** transmission → (↑) **Nissan:** — **Isuzu:** —	**Nissan:** as in Thai. **Isuzu:** —	

Source: JETRO 2000.3: *Nikei oyobi O-bei Jidosha Meka- no Ajia Semritsu* (in Japanese)

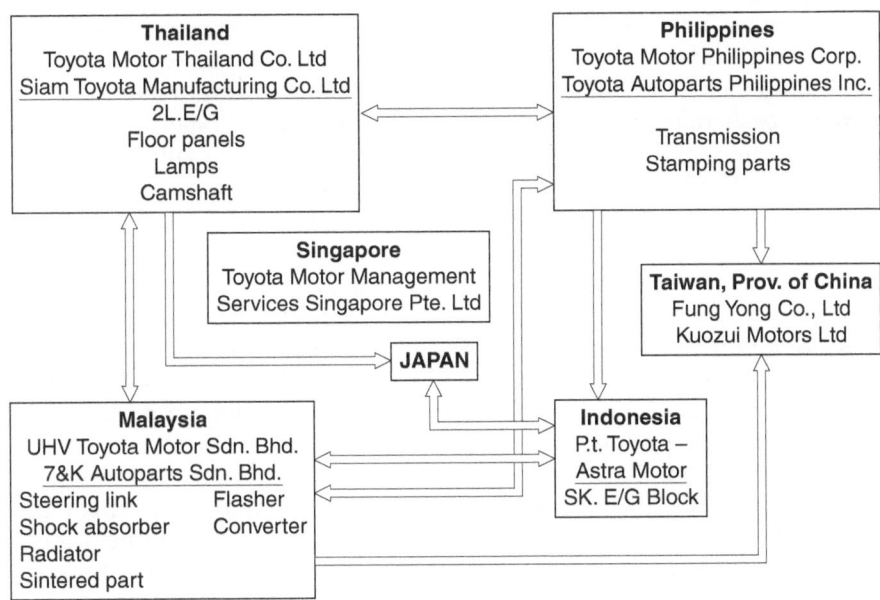

Figure 5.3 Toyota's parts and components ASEAN network
Source: UNCTAD (1996)

Reasons for the limited success of the Car Development Program

The government's basic rationale for the adoption of a local content policy was that compelling foreign assemblers to source parts and components from the domestic economy would induce the emergence of Filipino supplier firms that would absorb the technology for parts manufacturing. As mentioned earlier, the government expected – quite simplistically – that, in the process, domestic sourcing would induce the 'upgrad[ing of] engineering and pro-duction skills and provide new technological know-how to the country's indus-trial sector' (quoted from a BOI internal document of 1973). Worth remembering, however, is that Japanese assembly firms that entered the United States in the 1980s were compelled by American policy makers to raise their domestic value added and were thus induced to source from local – that is, American – firms (Okamoto 1999). In Europe, Nissan Bluebird cars manu-factured in the United Kingdom were barred from taking advantage of regional free trade unless the Japanese manufacturer raised domestic value added on the finished car (Pelkmans 1997). The presupposition in both cases was that a domestic car parts industry already existed, and that the Japanese transplants should merely be compelled to source from them. However, in the Philippine case, supplier firms, especially at the beginning stages of the programmes to develop the car industry, were close to being non-existent and those that did exist were not able to meet the quality–cost–delivery (QCD) requirements of the Japanese assemblers. The government was thus, in a sense, forced to provide a number of loopholes to the 'local content rule', allowing for its circumvention and eventually leading to the limited success of the programmes in terms of inducing technological transfer and upgrading.

A number of reasons could be cited as to why MNEs preferred not to out-source components manufacturing to local Filipino firms. First, there is generally a basic reluctance to transfer proprietary technologies to potential competitors among Filipino-owned firms, assuming these could be found. Such technologies have been developed after long years of experience and have incurred substantial R&D costs. Moreover, the production of some of these main components have been kept 'close to the heart' of the foreign assembler, because they were considered the core of their competitive advantage (e.g. gearboxes and engine designs). Thus the assemblers chose to produce them initially in-house and, later on, via majority-owned subsidiaries (such as Toyota Autoparts Phil., Honda Engine Mfg Phil., Honda Parts Mfg Corp. and Isuzu Autoparts Mfg Corp.) or by their first-tier suppliers, often but not always a part of the tightly knit *keiretsu*. This, in turn, reduced the need to transfer technology to local Filipino firms.

Second, there was the near-absence of local firms that could meet the standards of the contractor–assembler in the short term. It could be argued (Doner 1987; Abrenica 2000a: 89) that by emphasizing local capability to produce major functional parts requiring rigorous technical specifications – such as engine blocks and transmissions – beyond the actual capability of local firms, government virtually forced the assemblers to integrate vertically instead of horizontally, as expected by policy makers.

Third, the formulation of the local content rule itself, as well as its actual implementation, ironically worked against the healthy development of the parts industry. Because of the government's desire to save foreign exchange in the process of import substitution, programme guidelines allowed local assem-blers to use a loophole while complying with the local content rule. This loop-hole was found in the definition of local content,[10] which allowed a firm to comply with the domestic content requirement by raising its export earnings rather than the value of locally manufactured components it used. It became rational for firms to choose to export because they could earn export credits (and hard cash in the process!), whereas the price of assembled cars rose as more local components were integrated into them (Abrenica 2000a: 89). According to Abrenica (2000a), estimates showed that increases in local content from 50 to 60 per cent would raise the price of a completed vehicle by 23.5 per cent; and a 32.4 per cent price increase would result from an up to 65 per cent rise in local content. That locally produced parts had higher prices than imported ones was due mainly to the inefficiency of their production.

Moreover, owing to a lack of technical expertise and manpower for moni-toring compliance, the BOI failed to enforce the local content requirement strictly. And yet, monitoring compliance was crucial to the success of the local content rule, because the cost penalty would have forced firms to limit the number of models and frequency of their changes, which in turn would have reduced the fragmentation of the parts market (Abrenica 2000a: 90).

Fourth, the adoption of policies that fragmented an already narrow market for cars was inconsistent with the desired success of the local content rule. In fact, right from the start, scale economies in both assembly and parts manu-facturing could not be achieved because of the inability to keep the number of car assemblers to the required minimum. With the eventual deregulation of the domestic market still protected by a 40 per cent tariff on CBUs, the pro-liferation of brands and models in the absence of part standardization meant that the imposition of a local content rule could only translate into inefficient production and, hence, inflated car prices that further dampened demand.

It must be admitted that there had been no systematic effort on the part of the government to encourage technology transfer *directly* and thus to promote linkages with the domestic economy. Only indirect means were used, such as the adoption of a protection policy and the imposition of a local content rule. In contrast, governments of other countries were more proactive in encouraging the creation and deepening of linkages between MNEs and local firms. For instance, Singapore has a Local Industry Upgrad-ing Programme (LIUP) to 'upgrade, strengthen and expand the pool of local suppliers to foreign affiliates' (UNCTAD 2001) that are efficient, reli-able and internationally competitive. Under the programme, organization and financial support are offered to upgrade and develop vendors. Foreign firms are encouraged to enter into long-term contracts with local suppliers, which are, in turn, assisted to upgrade their products and processes. In other words, Singapore has combined a targeted FDI promotion strategy with a linkage programme. In the Philippines, technology upgrading of local firms, which is the key determinant of their ability to qualify as suppliers to interna-tionally competitive firms, had not been vigorously pursued by the govern-ment. While some government agencies might have been mandated to assist private firms in improving their technology, the actual operations of these agencies have been hampered by a lack of financing and qualified personnel. Moreover, although the Philippine government, through the BOI, attempted to introduce information provision and matchmaking that could help domestic firms, particularly SMEs, to link up with foreign affiliates, this was done on a very limited scale and lacked long-term sustainability.

Thus the failure of public policy to encourage technology transfer directly through the promotion of linkages between FDI and the domestic economy meant that private firms were on their own to determine the degree of trans-fer of know-how to local firms, while working within the political and policy framework provided by the state. In the following section, I inquire into the degree of technology transfer in Japanese transplants in the Philippines, given the policy environment in which they had to operate.

Some indicators of technology transfer in Japanese transplants

The process of technology transfer passes through different stages: the identifi-cation, or search, stage; the transaction stage; and the internalization stage (Wong 1991: 10).[11] In what follows, I shall assume that the process had gone through the first and second stages, so that we refer to the third, or internaliza-tion, stage when speaking of technology transfer. There are two modes by which the transfer can take place, namely through international transfer or through domestic diffusion (Wong 1991: 100–111). In the first case, where technology is developed in another nation, transfer occurs through the MNE's establishment of some parts of its value-chain activities locally (that is, through FDI), through commercialized 'contractual' means, i.e. through trade of goods or of a package of goods and services (such as outright purchase of technology, licensing, technical/consultancy services, franchising, etc.), or through 'non-commercial' means, such as through technical assistance between governments, publications, international migration, overseas training, etc. On the other hand, with domestic diffusion, technology is transferred through a backward linkage relationship, as when MNEs source from local suppliers; forward linkage, as from MNEs to local distributors or end-user industries; and through movement of trained staff from MNEs to local firms. Wong (1991), in his study of technology transfer and absorption via the subcontracting route between MNEs and supplier firms in Singapore, identified four ways by which MNEs can affect the rate of technological learning in the SME suppliers. These were:

1 *direct transfer of know-how*, wherein the MNE commits resources in a conscious effort to transfer certain technology at its disposal;
2 *learning facilitation*, wherein no actual transfer of know-how takes place, but feedback obtained from the MNEs, with regard to a stringent quality/performance assurance control system over supplied output, *facilitates* the recipient's technological learning;
3 *indirect technology transfer*, wherein in some ways unintended by the MNEs – hence involving little or no resource commitment on the part of the technology source – the subcontracting relationship itself imparts certain technological knowledge to the subcontractor;
4 *inducement*, wherein the subcontractor is induced by the relationship to commit certain technological investments that he would not undertake in the absence of the subcontracting relationship.

International technology transfer

When the international transfer mode is used – that is, through the local establishment of the MNE's value chain – direct transfer of know-how to production engineers, technicians and production workers can be expected to take place. My interviews with managers of assembly and autoparts manufacturing plants that are either wholly owned or majority-owned by the Japanese parent firm[12] indicated that, generally speaking, production technology is directly transferred. Although the beginning stages required much input in terms of expatriate time and technical supervision, current practice in most transplants is for production management to be handled directly by a Filipino worker, even if the title of Production Manager may still be held by a Japanese worker. This indicates that, in most cases, operations technology has been fully transferred. Together with production technology comes quality control, which is likewise handled by Filipino engineers and technicians. Transfer of knowledge is undertaken through manuals (usually written in English), on-the-job training (OJT), seminars and other training means. Often, when a new technology has to be learned, or in the case of a full change in model, local engineers and/or technical people (supervisory level) are sent to a Japanese plant for training and interaction with Japanese engineers and line workers. In some cases, line supervisors and one or two production workers are sent to Japan for OJT. Then these trained technical staff (including chosen line operators who eventually become supervisors) 'echo' the training they received to the rest of the production workers in the local plant. In the case of at least one autoparts assembly plant, the manual for a full model change – for use both in-house and in plants located in English-speaking countries (such as India, Pakistan and New Zealand) – was written up in the Philippines. In most cases, and especially at the beginning stages, Japanese engineers/technical people were dispatched to the local plant to conduct the training directly, to supervise the local trainers, or to be around for troubleshooting. Over time, however, it has been found to be more economical to send Filipino technical people and supervisors to Japan for a week or so of training.

However, we have not found a case where the transfer of *design technology and new product development* has taken place. Interviews of plant managers showed that indeed design technology and product development is kept the domain of the parent firm in Japan. Of course, information on possible technology design changes and on new products may be supplied by the local affiliate, as when its personnel pick up information in the course of their contact with buyers. But the actual decisions on whether or not to embark in this direction are taken by the parent firm. Full model changes are likewise decided at the parent-firm level where R&D activity is centralized.

Although design technology is generally absent in local affiliates, they may undertake adaptation to local conditions of designs which invariably originate from the parent firm. For instance, a horn assembly plant associated with a German multinational was given permission to adapt a parent-designed horn to Philippine conditions, where drivers blow their horns much more often than in Europe or where horn connections may be submerged in water during floods. Such conditions necessitated more revolutions and sealing of the horn to solve these problems, leading to higher cost. Such adaptations were studied and suggested by the local plant engineers, and given a seal of approval by the parent firm. In other cases, including in Japanese transplants, simple modification of machine design has been undertaken by the affiliate firm. Still other cases needing little intervention from the parent firm are those concerning variations on models, since full model changes take place usually over a five-year cycle. On the other hand, without graduating to design technology and new product development, Filipino engineers cannot claim to have fully absorbed the technology.

This finding is confirmed by many studies and, most recently, by Urata (1999), who analysed both the patterns of technology transfer through FDI by Japanese MNEs and the determinants of successful technology transfer. Using data from a survey of about 133 firms in nine Asian countries operating in four industries (general machinery, electrical machinery, transport machinery and precision instruments) conducted by the Nikkei Research Institute of Industry and Markets in 1991, he found that a large part of manufacturing technologies, such as operational technologies, maintenance and inspection technologies, had been transferred to Asian affiliates of Japanese firms. However, sophisticated technologies – such as design technology and development of new products – had not been transferred in many of the firms surveyed. Moreover, successful technology transfer[13] in manufacturing technologies – but not in sophisticated technologies – took place through the realization of local ownership. He also found that the transfer of sophisticated technologies was very slow in Japanese firms that had been set up to take advantage of cheap labour. This means that export-oriented Japanese firms in the export-processing zones cannot be expected to transfer sophisticated technology. In the case of automotive firms, technology can be said to be fairly standardized, and the assembly of CKD packs, especially, is usually of a 'screwdriver' type.

Technology transfer through domestic diffusion

It had been noted earlier that because the Japanese assembly firms had opted for in-house production or, whenever possible under the local content rule, through CKD imports, little subcontracting took place, unlike in Japan. Table 5.5 shows that in 1994, subcontracting (represented by the ratio of work done by others to total output of a manufacturing industry) was below 10 per cent for the manufacturing sector in general. Automotive assembly, with a 28 per cent subcontracting ratio, already had the highest ratio among manufacturing industries. Parts and component manufacturing, in contrast, recorded only a 1.29 per cent subcontracting ratio. To what extent has technology been transferred by Japanese transplants to local firms through the second mode – that is, through backward linkage, through either subcontracting or outsourcing?

Table 5.5 Ratio of work done by others to total output of selected manufacturing industries with 10 per cent or more foreign equity, 1994

PSIC	Industry	20–99 workers	100 or more workers
	Total manufacturing	7.53	7.60
3222	Ready-made clothing	23.96	7.13
3229	Wearing apparel	15.66	12.10
3514	Pesticides, fungicides, etc.	—	18.72
3522	Drugs and medicine	10.37	17.11
3523	Soap and cleaning preparations	8.18	21.43
3560	Plastic products, n.e.c.	10.5	19.39
3610	Pottery, china, earthenware	—	17.78
3692	Structural concrete products	12.86	24.84
3822	Agricultural machinery and equipment	14.10	—
3824	Special industrial machinery and equipment	15.80	—
3836	Electric wires and wiring devices	—	13.28
3839	Electrical apparatus and supplies	16.59	19.02
3843	Motor vehicles	—	27.88
3845	Vehicle parts and accessories	—	1.29
3852	Photographic and optical instruments	6.89	13.22

Source: NSO file (1994)

Table 5.6 seems to indicate that very little transfer has been taking place. In all except one response to the interview question on whether technology transfer takes place to supplier firms, the answers were negative. In the case of the two car assembly firms in the survey, the reason for the negative response was that they do not possess the technology being applied by the supplier firm. These firms were chosen by the parent firm in Japan for their track record of reliability in quality, cost and delivery (QCD), and hence were knowledgeable about their output technology. And even if the technology were available to the parent firm, it would normally not be its core competence, so that if necessary, the parent firm would contact the parent firm of the local supplier in Japan in order to provide the latter with technical assistance. This underscores a major fact in the Philippine automotive industry, namely, that Japanese transplants would most likely be dealing with firms affiliated to *Japanese* parent firms, whether or not they were part of the assembler's *keiretsu*.

Among the component manufacturers, the reasons for their negative responses generally resembled those given by the assembly firms – namely, that they did not possess the technology being applied by the supplier firm, that the supplier firm was more knowledgeable than they about the relevant technology, or that they relied more on imports for their raw materials – so that little or no technology transfer to the local supplier firm was required. The only affirmative answer, however, provided a reason that was almost tailor-made to the expectation of policy makers: the component manufacturer provided assistance with the technology of moulds for the making of rubber parts and with precision manufacturing, in the absence of any formal technical assistance agreement. Moreover, the sourcing firm also indicated that it had provided learning facilitation technology through its demands for raw material and price reduction.

Where there is no direct transfer of production technology or know-how, it is still possible to have transfer in the form of *learning facilitation* – that is, Filipino firms could be learning from the presence of Japanese transplants through product specification and exposure to efficient manufacturing processes. In dealing with Japanese buyer firms, they are 'forced' to learn how to keep up with their QCD requirements. This may be done through initial and periodic scrutiny of their production workplaces and practices, if only to assure the Japanese buyer that they are up to par with their requirements. MNEs' requiring supplier firms to achieve ISO certification or some form of certification of productivity enhancement is an important development. It implies that simply being a supplier to a Japanese MNE means that one will have to show measurable indicators of productivity growth.

Just how important learning facilitation is was revealed in a study on technology transfer through subcontracting in the electronics industry of Singapore. From a survey of SMEs subcontracting with three Japanese, three American and two European firms, Wong (1991) found that overall direct transfer activities appeared to be less important for SME suppliers than other categories of indirect transfer. Learning facilitation through quality testing and diagnostics feedback, together with spillover transfer through product specification and exposure to general 'good manufacturing practices' due to transactions with MNEs, appeared high in perceived importance to them.

Table 5.6 Technology transfer to suppliers by Japanese transplants in the Philippines

Auto firm	Product/customer	Q: Do you provide technology assistance to your suppliers?	Reason/notes
Car assembler I	Car assembly	No	Many parts suppliers are independent Japanese firms, joint ventures, or foreign ones with their own technology. We choose them on the basis of their past performance of quality control, price reduction and Just-in-Time management. They determine the management of their production system.
Car assembler II	Car assembly	No	The parts our firm obtains from local firms are determined by our parent firms. Thus, our firm's technological know-how to supply or teach them is weak/absent. If there is a need, our parent firm can request the parent firm of the local supplier to provide the guidance.
Car assembler II Autoparts 1	Transmission; drive shaft/Toyota Motors (esp. ASEAN, Japan, South Asia)	No	Assistance and supply of technology come from our parent firm in Japan.
Supplier 2	Switches/all	No	
Supplier 3	Engine parts; metal parts/ Honda	No	No local firms can supply firm's needs. Firm's supply of basic materials, which are aluminium ingots and steel, are imported from Japan.
Supplier 4	Metal; car air-conditioning	No	Supply from local firm (Japanese) is very small.
Supplier 5	Glass for autos	No	Our material supplier firms do not need technical assistance. They possess the technology they need.
Supplier 6	Autoparts; airbags	No answer	No answer.
Supplier 7	Automatic wiring harness	No	Interacts with our parent firm.
Supplier 8	Wiring harness	Yes	Technology for moulds, such as for rubber parts, technology for precision manufacturing, technology for raw material reduction, price reduction. However, no formal TA is provided.
Supplier 9	Assembly of car stereo; assembly of CD-ROM	No	Raw materials and parts for assembly are imported; hence, no need for technology transfer.
Supplier 10	Diecast	No	Our firm does not have the technology needed for the major materials we use, such as aluminium and zinc ingots.

Source: Interview survey, Institute of Developing Economies (1999)

Thus the only direct form of transfer Wong found from the case studies – and this ranked as moderately important – was advice or training on quality management systems and other 'good manufacturing practices'. He then concluded that in Singapore, the main form of direct technology transfer is of 'soft' rather than 'hard' technologies or specialized know-how of production. This also seems to be generally the case in the Philippines.

In addition, there has been a low turnover rate of workers and engineers – who usually quit for more lucrative jobs in the Middle East – possibly because of the limited size of the market. The only big case of labour movement in recent history was the establishment of an assembly plant by another MNE (Ford). Thus the spillover effect of technology learning via labour movement may be considered quite limited.

Concluding remarks

It must be conceded that, other than the general allusion to the policy objective of technology transfer in the automotive industry development programmes of the Philippine government, there is no systematic government policy that targets the achievement of technology transfer in the industry. It had been simply assumed that the application of high tariff protection and local content requirement would be enough to attract foreign assembly firms to invest in the country and, in the process, transfer their technology to Filipino supplier firms. While it is true that the import substitution policy did manage to attract foreign MNEs to invest in assembly, their activity of transferring technology had been limited to in-house production or that in directly controlled subsidiary firms, and these mainly in the field of operations technology, inspection and quality control, but not in design technology, let alone in the development of new products. Eventually, when policy increasingly urged the stimulation of local production capabilities for parts and components, the foreign assembly firms' response was to persuade their suppliers in Japan to invest in the Philippines. But even in their subcontracting or outsourcing relations with these supplier firms, there is no strong indication that Japanese assembly firms directly transferred production technology. One important reason is that the supplier's technology is usually not part of the assembler's core competence. Supplier firms have been chosen on the basis of their track record of reliable supply, so that the assemblers would not have to bother about teaching them the necessary technology. However, even when there is no direct transfer of production know-how, the transfer of learning facilitation technology is definitely taking place: supplier firms are getting exposure to the Japanese product specification and good manufacturing practices.

In the face of a more liberal market situation in both the Philippines and the region generally, what are the prospects for a greater amount of technology transfer taking place in the industry? By 2004, the Philippines is expected to adopt a uniform tariff policy of 5 per cent for both CBUs and imported parts and components. By itself, this policy may make parts production even more unattractive. Moreover, the local content rule – which was supposed to be abolished in 2000 in compliance with the TRIMs Agreement under the GATT–WTO and for which the Philippines was recently given a reprieve as a result of the Asian financial crisis – will sooner or later disappear. With the eventual demise of protection policy, will the production of car parts remain viable? Will MNEs continue producing, say, transmissions, servo brakes and gearboxes in the absence of a local content rule? The inflow of FDI into the industry in the 1990s, in spite of the prospects of abolition of the local content rule, may merely be a short-term response to policy. Even the exporting of major components may merely have been occasioned by MNEs' attempts to comply with policy in order to acquire a foothold in the domestic market.

In spite of more than three decades of application of import substitution policy for development of a local automotive industry, there is no such local car assembler to speak of, nor is there a supply network extensive enough to attract foreign assemblers once the trade barriers are dismantled. While the extent of 'upgrad[ing of] engineering and production skills' as well as the provision of 'new technological know-how to the country's industrial sector' (BOI internal document, 1973) via the development of assembly and parts production remains largely unknown, the burden on domestic consumers of three decades of high automobile prices is certain (Aldaba 2000; Abrenica 2000b). The government has thus finally decided to 'bite the bullet' in liberalizing the market, albeit gradually in order to soft-pedal the crash of basically uneconomic ventures.

A more open economic environment promises to bring out the competitive strengths of the country. Already, car assemblers are gearing up for the eventual free trade in the region, with the implementation of the AFTA. Ford is rumoured to have assigned small car production to its Philippine plant, while Japanese firms are eyeing the region for the production of specific parts (e.g. Toyota and Mitsubishi in transmissions). Analysts who have studied the direction of the global automotive industry closely are convinced that 'upgrading the local autoparts industry should be at the core of the policy of any Asian government' (Veloso 2000: 37). With the new industry developments in the direction of micro-based technologies and Internet-based supply-chain management, the Philippines may still hope to find a few niches in electronic parts manufacturing, as well as in the provision of e-services. However, taking this direction will require both government and private firms to think hard and deeply, and to act decisively and cooperatively in creating globally competitive suppliers out of local firms.

Notes

1 Moreover, the demand by developing countries for more capital inflows through foreign direct investment (FDI) met the search for labour-abundant production sites launched by automotive firms based in developed countries that would help them maintain their competitive edge in the strongly competitive world market for cars.

2 Given the small domestic demand for automotive vehicles at the time, the chosen number of five was considered by critics as being too large to achieve scale economies in production. It is said that the original plan provided for only two firms, but that political accommodation allowed five MNEs to be admitted,

namely Delta Motors Corp. (owned by Richard Silverio Jr with Japanese equity by Toyota), Ford Inc. Philippines (Ford-American); Yutivo-Francisco Motors Corp. (General Motors Pilipinas Inc. with Japanese equity GM-Isuzu), DMG Inc. with German equity (Volkswagen) and Chrysler Philippines (or Canlubang Automotive Resources Corp., with Japanese equity from Mitsubishi).

3 Examples of such technology transfer were found in subcontracting torque converters to Aichi Kogyo, wheels to Chuo Seiki, engine valves to Aisan Kogyo, driving parts to Aishin Seiki and brakes to Hosei Brake (Nishiguchi 1994: 107).

4 Together with the PCMP was the Progressive Motorcycle Manufacturing Program (PMMP) and the Progressive Truck Manufacturing Program (PTMP), which similarly protected domestic production by high tariffs but required a local content. With the new MVDP, the two programs were similarly modified into the Motorcycle Development Program (MDP) to replace the PMMP and the Commercial Vehicle Development Program (CVDP) to replace the PTMP.

5 The new formula for calculating local content was supposed to shift emphasis to parts manufacturing and away from foreign exchange generation, as follows: Local content = (points × local content of components) +15 per cent assembly allowance. 'Points' refers to the ratio of CKD price of individual parts of components to the CKD full pack price; local content of components refers to the selling price of each component less imported materials, depreciation of imported equipment and other foreign costs; assembly allowance refers to other local materials and supplies used in assembly. A 15 per cent cost penalty allowance on local parts means that the selling price of the local component used in the computation may be, at most, 15 per cent higher than the landed cost of the part taken out of the imported package.

6 In the meantime, the credits for non-automotive parts exports would be gradually phased out until 1993.

7 The main category of cars refers to those with engine displacement of 1,200–2,800 cc, as against the People's Car category of less than 1,200 cc displacement.

8 These were for cars with engine displacement of 2,190 cc or greater for gasoline-fed, or 3,100 cc or greater for diesel-fed.

9 The depressed state of parts imports, having failed to recover from the 1997 Asian financial crisis, may, of course, be mainly responsible for such net trade phenomena.

10 Local content ratio = FOB import price of local parts + foreign exchange earnings on imports/FOB export cost of CBU.

11 Wong (1991) identified three stages through which the process of technology transfer takes place, namely: (1) the identification, or search, stage, wherein the recipient searches out and assesses the potential technology sources for accumulation (or vice versa, meaning the technology sources search for the recipient); (2) the transaction stage, wherein the technology acquisition mechanism is negotiated and implemented; and (3) the internationalization stage, wherein the acquired technology is actually assimilated within the recipient organization.

12 Assembly and autoparts manufacturing plant managers mainly in the Calabarzon area (such as Honda Autoparts, Toyota Autoparts) were interviewed in June and September 2001.

13 Technology transfer was considered by the author as complete or successful when the response to the question of who is in charge of specified technological activities was 'local workers', and considered incomplete or unsuccessful if the answer was 'Japanese workers'.

References and further reading

Abrenica, M.J. (1998) 'The Asian Automotive Industry: Assessing the Roles of State and Market in the Age of Global Competition', *Asian-Pacific Economic Literature*, 12 (1) (May): 12–25.

—— (2000a) 'The Philippine Automotive Industry', paper for the PECC Conference, Manila, August (unpublished).

—— (2000b) 'Liberalizing the ASEAN Automotive Market: Impact Assessment', paper for the PECC Conference, Manila, August (unpublished).

Aldaba, R. (2000) 'Increasing Globalization and AFTA in 2003: What Are the Prospects for the Philippine Automotive Industry?', discussion paper, Series 2000–42, Makati: Philippine Institute for Development Studies.

Doner, R.F. (1987) 'Domestic Coalitions and Japanese Auto Firms in Southeast Asia: A Comparative Bargaining Study', unpublished Ph.D. dissertation, University of California, Berkeley.

Findlay, C. and Abrenica, M.J. (2000) 'The ASEAN Automotive Industry: Challenges and Opportunities', paper for the PECC Conference, Manila, August (unpublished).

Glass, A.J. and Saggi, K. (1998) 'International Technology Transfer and the Technology Gap', *Journal of Development Economics*, 55: 369–398.

Hatch, W. (1998) 'Grounding Asia's Flying Geese: The Costs of Depending Heavily on Japanese Capital and Technology', Seattle: National Bureau of Asian Research. Online. Available at http://www.nbr.org (*NBR Briefing*, April).

Hatch, W. and Yamamura, K. (1996) *Asia in Japan's Embrace*, Cambridge: Cambridge University Press.

Lall, S. (1977) 'Technology Transfer by Multinational Firms: The Resource Cost of Transferring Technological Know-how', *Economic Journal*, 87: 242–261.

—— (2000) 'Foreign Direct Investment, Technology Development and Competitiveness: Conceptual Issues and Empirical Review', draft version for the study *Foreign Direct Investment, Technology and Competitiveness in East Asia*, World Bank Institute (unpublished draft, October).

Nipon, P. and Wangdee, C. (2000) 'The Impact of Technological Change and Corporate Reorganization in the ASEAN Automotive Industry', paper for the PECC Conference, Manila, August (unpublished).

Nishiguchi, T. (1994) *Strategic Industrial Sourcing: The Japanese Advantage*, New York: Oxford University Press.

Okamoto, Y. (1999) 'Multinationals, Employment Creation and Spillover Effects: The Case of the Philippines', *Philippine Review of Economics and Business* (December): 177–204.

Pelkmans, J. (1997) *European Integration: Methods and Economic Analysis*, Hong Kong: Longman Asia.

Saggi, K. (2000) 'Trade, Foreign Direct Investment, and International Technology Transfer: A Survey', Policy Research Working Paper 2349, Washington, DC: World Bank (May).

UNCTAD (1996) *World Investment Report 1996*, Geneva: United Nations.

—— (2001) *World Investment Report 2001*, Geneva: United Nations.

Urata, S. (1999) 'Intrafirm Technology Transfer by Japanese Multinationals in Asia', in Encarnation, D.J. (ed.) *Japanese Multinationals in Asia: Regional Operations in Comparative Perspective*, New York: Oxford University Press.

Veloso, F. (2002) 'The Automotive Supply Chain: Global Trends and Asian Perspectives', paper prepared for RETA 5875, 'International Competitiveness of Asian Economies: A Cross-Country Study', Manila: Asian Development Bank. Online. Available at http://www.adb.org/Documents/ERD/Working_Papers/wp003.pdf.

Wong, P.K. (1991) *Technological Development through Subcontracting Linkages: A Case Study*, Tokyo: Asian Productivity Organization.

6 Indonesian industrial policy in the automobile sector

Focus on technology transfer

Lepi T. Tarmidi

This chapter begins with a quantitative overview of the development of the Indonesian automobile industry, while the second section offers an analysis of the most important governmental policies for the industry, with particular emphasis on the question of technology transfer. The fourth section investigates the role Japanese producers in Indonesia play in technology transfer.

Development of the Indonesian automobile industry

Automobile assembling in Indonesia already existed in 1927 when General Motors built a plant in Jakarta, but it was not developed further and GM remained the only car assembler for many years. Cars were imported and, until the 1960s, the number was still relatively small. Until 1969, imports of automobiles were not regulated and the industry consisted mainly of trading activities with very limited assembling operations (Shauki 1996: 3). However, assembling activities started developing in 1964 with semi-knocked-down components, but then the process progressed into the assembling of knocked-down parts in 1971. Automobile assembly production has tended to undergo, big fluctuations: it increased sharply from 22,118 units in 1972 to 212,669 units in 1981, but then dropped drastically in 1985 to 139,438 units. In 1990, production increased again sharply to 271,712 units, but then dropped to 172,234 units in 1992, only to increase rapidly again to 387,541 units in 1995. But owing to the financial crisis that started in mid-1997, automobile assembly production dropped drastically to only 58,079 units in 1998, but recovered to 292,710 units in 2000. In 2001, production declined slightly to 279,187 units because the government began to allow imports of built-up automobiles, which reached 16,617 units that year. The year 2002 saw some recovery as sales of automobiles from January to August reached 212,378 units.

Automobiles assembled in Indonesia are categorized into sedans for personal use and cars for commercial use. The majority of automobiles produced in Indonesia belong to the commercial car category. In 1972, the proportion of commercial cars was 15,993 units, or 72.3 per cent of total automobile production, versus sedans at 6,125 units, or 27.7 per cent. For 1985, the figures were respectively 115,239 units for commercial cars, or 82.6 per cent, versus only 24,199 units of sedans, or 17.4 per cent. In 1996, the proportion changed in favour of commercial vehicles, with respectively 290,191 units versus 35,304 units of passenger cars (89.2 versus 10.8 per cent), while in 2001, the proportion did not change much: 246,950 versus 32,237 units, or, respectively, 88.5 versus 11.5 per cent.

In 1995, there were as many as thirty-two automobile brands represented in Indonesia, with eighteen assembly firms achieving a total production capacity of 547,622 units per annum. But of these eighteen firms, only fourteen were active, producing twenty car brands and a production capacity of 514,622 units per annum. In 1998, the number of car brands fell to twenty-two, but production capacity increased substantially to 750,000 units per annum. Japanese brands easily dominated automobile production in Indonesia. In the sedan category, 75.4 per cent of car production in 1983 was Japanese. But sales of Japanese brands in the sedan category declined relatively sharply between 1991 and 1996 from 34,781 units (76.2 per cent) to 21,445 units (48.9 per cent). Japanese passenger cars soon regained their market share position, though not fully, to 70 per cent in 2001. In the commercial vehicle category, Japanese domination was almost complete. Of all commercial vehicles produced in 1983, 91 per cent were Japanese brands. The market share percentage increased to 94.9 per cent in 1996 of all commercial vehicle sales, but declined slightly to 92.8 per cent in 2001 (see Table 6.1). By 2001, there were at least twenty-five automobile brands being assembled in Indonesia – ten from Europe, eight from Japan, four from Korea and three from the United States – by nineteen automobile assembly firms (Gaikindo 2001: 38–40; Ministry of Industry and Trade). Meanwhile, Citroën, Renault, Fiat and Holden have stopped production in Indonesia, owing to the small and declining demand for these brands. On the other hand, new brands such as Hyundai, Daewoo, KIA/Timor and Jaguar are invading the market.

A way to assess the extent of technology transfer in the automobile assembly industry is to look at the development of automobile component production and exports in the country. Logic dictates that the higher the local content proportion in the final product of a car, the greater the success of the technology transfer from abroad. In developing countries like Indonesia, there is practically no research and development (R&D) in automotive technologies, and hence developing countries depend entirely on foreign car producers and foreign technology. Therefore, the only way to develop an automobile industry in developing countries is through technology transfer.

The Indonesian automotive component industry has grown rapidly as a result of the deletion programme pursued by the government since the early 1980s, with the goal of increasing the portion of local content in car assembling. Though the automotive component manufacturers in Indonesia were much more numerous (223) than those in the Philippines (154) in 1996, they were still far behind those in Thailand (488) and Malaysia (515). In Thailand, the highest concentration of firms was to be found in the body parts manufacturing industry (100), followed by that of engine component manufacturing (106). In Malaysia, it was also in the body parts manufacturing industry (137), followed by drive system/tyre wheel manufacturing (100).

Table 6.1 Sales of motor vehicles in Indonesia by brand in 1991, 1996 and 2001

	1991	1996	2001
Sedans			
Toyota	10,678 (23.4)	8,612 (19.6)	12,053 (34.1)
Suzuki	3,479 (7.6)	2,534 (5.8)	4,558 (12.9)
Daihatsu	4,285 (9.4)	205 (0.5)	0
Mitsubishi	3,732 (8.2)	1,993 (4.5)	1,338 (3.8)
Honda	8,205 (18.0)	6,357 (14.5)	6,656 (18.9)
Mazda	4,402 (9.6)	744 (1.7)	3
Nissan	0	1,000 (2.3)	91 (0.3)
Subtotal	34,781 (76.2)	21,445 (48.9)	24,699 (70.0)
Ford	2,242 (4.9)	3,651 (8.3)	54 (0.2)
BMW	3,131 (6.9)	3,788 (8.6)	2,830 (8.0)
Mercedes	1,223 (2.7)	3,829 (8.7)	1,690 (4.8)
Hyundai	—	2,056 (4.7)	2,959 (8.4)
Timor	—	6,042 (13.8)	2,091 (5.9)
Volvo			131 (0.4)
Peugeot			31 (0.1)
Renault			31 (0.1)
KIA			112 (0.3)
Others	4,314 (9.4)	3,103 (7.1)	669 (1.9)
Total	45,691 (100)	43,914 (100)	35,297 (100)
	[17.4]	[13.0]	[11.8]
Commercial			
Toyota	65,462 (30.1)	67,047 (22.9)	68,016 (25.7)
Suzuki	39,138 (18.0)	55,869 (19.0)	48,629 (18.4)
Daihatsu	39,491 (18.2)	38,438 (13.1)	20,592 (7.8)
Mitsubishi	47,698 (21.9)	74,261 (25.3)	64,766 (24.5)
Isuzu	16,538 (7.6)	42,967 (14.6)	31,290 (11.8)
Hino			3,035 (1.1)
Nissan			3,924 (1.5)
Honda			4,854 (1.8)
Mazda			240 (0.1)
Subtotal	208,327 (95.8)	278,582 (94.9)	245,346 (92.8)
Peugeot			1,855 (0.7)
Mercedes			1,316 (0.5)
GM			2,383 (0.9)
Chevrolet			531 (0.2)
Hyundai			5,895 (2.2)
KIA			6,422 (2.4)
Others	9,055 (4.2)	14,903 (5.1)	528 (0.2)
Total	217,382 (100)	293,485 (100)	264,276 (100)
	[82.6]	[87.0]	[88.2]
Grand total	263,073	337,399	299,573
	[100]	[100]	[100]

Source: Gaikindo (2001)

Note
Numbers in ( ) and [ ] indicate percentages

In Indonesia, the distribution of firms was almost equal among engine component manufacturing (51), body electrical component manufacturing (47) and body parts manufacturing (40). The smaller number of automotive component manufacturers in Indonesia as compared with those in Malaysia or Thailand was not the cause of the later development of the industry in Indonesia, which rather was due to government policies, inflow of foreign investment and local potential manufacturers' capabilities to respond (Sato 2001: 3–8, 9, 10; see Table 6.2).

Indonesian industrial policy in the automobile sector

The Indonesian government issued a decree in 1969 regulating imports of motor vehicles, both in completely built up (CBU) and in completely knocked down (CKD) condition, by sole agents or brand-owners of Indonesian nationality. Consequently, these importers had a monopoly over imports of certain automobile brands. In March 1971, the government banned imports of CBU cars in order to protect local assembly firms. During the first stage of import substitution policy, the ownership of assembly firms and sole agents was reserved purely for domestic capital by setting an entry ban on foreign automobile makers, and only licensing agreements were allowed. A certain automobile brand name was to be produced and distributed under a single-ownership firm.

The Indonesian government has always cherished the ambition to establish a national automobile industry through import substitution. To increase the role of local manufacturing, the government stipulated a deletion programme in 1979. It encouraged the local manufacture of automotive components and aimed at increasing the local content portion, leading in the end to full manufacturing by 1986. The Ministry of Industry also issued a regulation in 1981, in which the number of brands and models were to be limited owing to the relatively small market size, so that economies of scale could be achieved (compare also Sato 2001: 3–20). The objective was to allow for a larger market share for each brand and hence increase efficiency

Table 6.2 Number of manufacturers of major automotive components in four ASEAN countries in 1996

Component specification	Indonesia	Thailand	Malaysia	Philippines
Engine component	51	106	79	25
Drive system/tyre wheel	26	75	100	30
Body part	40	140	137	33
Body electrical component	47	78	90	26
Exhaust system	10	11	16	9
Fuel system	17	31	31	2
Brake system	20	28	36	20
Suspension system	12	19	26	9
Total	223	488	515	154

Source: Sato (2001: 3–8)

and lower production costs. However, like many policy measures in Indonesia, the policy was never effective in limiting the number of automobile brands, owing to a lack of law enforcement and to the many existing vested interests in the industry. On the contrary, the number of car brands increased rapidly from year to year, crowding the expanding but still relatively small domestic market.

Next, in a 1983 stipulation, the government set a time schedule for each kind of automotive component to be manufactured locally. Studies by Indonesian and Japanese experts have pointed out that the programme to stimulate local production of compulsory machinery failed, to a large extent because the small and medium-scale manufacturers faced many problems relating to the lack of technology, capital, and skills in technical areas, marketing and management (Sato 2001: 2–19). However, the growth of the automotive component industry has been substantially accelerated, partly by the deepening policy in support of locally based industry (deletion programme) for four-wheel and two-wheel motor vehicle components since 1979 and 1980, and partly by the foreign and domestic investment boom after 1988 (Sato 1998: 111). 'Over the years as the mandatory deletion program proceeded, the imported contents in the assembly industry have fallen on the one hand, and those of the automobile and motorcycle industry have risen on the other hand' (ibid.). Of the total material in the automobile component industry, as much as 89 per cent came to be imported, which was also the case for as much as 58 per cent in the motorcycle component industry (ibid.: 119).

The government's original idea was to induce assemblers to procure locally made components from outside firm suppliers rather than through in-house production. However, from the viewpoint of foreign principals, there were very few local component manufacturers that were reliable enough to guarantee stable supply in large quantities and that could meet the required quality standard and guarantee timely delivery. Faced with the time schedule of the deletion programme, most Japanese principals persuaded their subcontractors and suppliers in Japan to undertake direct investments in Indonesia or to license their technologies to Indonesians. For engines and rear bodies, which require large investment but have a small production volume, owing to the specific nature of the brand or model, the Japanese principals themselves invested in engine and body press factories. For other functional components, such as transmissions and brake systems, the government approved the establishment of only one locally owned factory, in order to secure production scale and local ownership. These core component manufacturers received foreign technical assistance (Sato 2001: 3–20).

In April 1992, the government issued a decree stipulating that foreign investments must have a divestment plan so that, after a certain time period, majority equity shares are transferred to Indonesian nationals. This policy can be seen partly as an effort by the government to transfer technology from foreigners to Indonesian nationals in an indirect way. However, in the June 1994 regulation this obligatory divestment policy was abandoned and full foreign ownership was once again allowed for the first time since the ban in 1974 (Tarmidi 1996: 2, 3).

Though the time schedule for the deletion programme had to be amended a couple of times, the programme continued up to 1993. However, under pressure of global trade liberalization, the Indonesian government eventually had to abandon non-tariff trade regulations, and the compulsory deletion programme in 1993. Manufacturers became free to determine the kind of component they wanted to produce, and it was not compulsory any more to manufacture a certain predetermined component attached to a rigid time schedule. The import ban on CBU passenger vehicles was also lifted and the deletion programme was subsequently substituted by a tariff incentive system, in which higher local content ratios were compensated with lower tariffs for CKD imports. The government provided an import duty exemption on CKD imports for minibuses if the local content portion was at least 40 per cent, and for sedans if it was at least 60 per cent. This policy motivated carmakers, in particular Japanese ones, to develop their own automobile component industry and thus boost the production of minibuses in order to benefit from the duty exemption and to sell relatively cheap minibuses in the market. Because of the Trade-Related Aspects of Investment Measures (TRIMs) Agreement and the WTO guidelines prohibiting all local content policies, the Ministry of Industry and Trade lifted the local content regulation in 1999 (Sato 2001: 3–19).

When the government banned imports of CBU cars in March 1971 in order to protect the domestic automobile assembly industry, it also levied high import tariff rates on imports of CKD components, making automobiles very expensive in Indonesia. Later, when the import ban on CBU cars was lifted, the government compensated with high import tariffs to protect local car assemblers. This policy was necessary to protect the local automobile assembly industries because imports of CKD parts were still subject to equally high import tariff rates. For the government, high tariff rates are a source of income, because the government can lower the import tariffs and still protect the domestic automobile assembly industry, as long as the import tariff rates for CKD components are lower than the import tariff rates for CBU cars. The tariff rate policy for imports of CKD vehicles and automotive components under the tariff incentive system is shown in Table 6.3. The maximum tariff for imports of components was 40 per cent, but if the local content proportion exceeded 40 per cent, the tariff rate became zero. The June 1995 regulation allowed imports of CBU cars, but import tariff rates were very high. For sedans, the rate was 200 per cent, and for commercial vehicles 105 per cent, while for vehicles for public utility purposes, it was set at only 5 per cent.

According to Faisal H. Basri, the government pursued the wrong policy by linking the tariff incentive system with the local content proportion level, because no country has ever succeeded in developing an automotive industry by way of increasing the local content proportion. On the contrary, some countries, such as Spain, Belgium and Mexico, have succeeded in developing their own automotive industry after they abandoned the local content requirement policy. The object is to increase the production of automobiles in order to create an attractive captive market for component production, then investors will come. A small market is not attractive for an automotive component manufacturer (Tarmidi 1998: 91–92).

Table 6.3 Tariff rates for imports of completely knocked down (CKD) automobiles and components, 1993 and 1995

Vehicle category	Year	Import tariff rates for CKD automobiles (%)				Tariff rates for automotive components (%)			
		LC < 20	20 ≤ LC ≤ 30	30 ≤ LC ≤ 40	LC > 40	LC < 20	20 ≤ LC ≤ 30	30 ≤ LC ≤ 40	LC > 40
Category I	1993	40	30	20	0	40	30	20	0
	1995	25	15	10	0	25	15	10	0
Category IV (jeep)	1993	40	30	20	0	40	30	20	0
	1995	25	15	10	0	25	15	10	0
Sedan	1993	100	80	60	40[a]	40	30	20	0
	1995	65	50	35	20[b]	25	15	10	0
Category II	1993	40	20	0[c]		40[d]	20[e]	0[f]	0
	1995	25	15	0		25	15	0	0
Category III	1993	40	20	0		40	20	0	0
	1995	25	15	0		25	15	0	0

Sources: Republic of Indonesia (1993, 1995a)

Notes

LC, local content portion

a For LC = 40–60%; for LC > 60%, the tariff rate was zero

b For LC = 40–50%; for LC = 50–60%, the tariff rate was 10%; for LC > 60%, the tariff rate was zero

c For LC > 30%

d For LC < 10%

e For LC = 10–20%

f For LC > 20%

In February 1996, Suharto, then President, signed a decree appointing the Timor Putra Nasional company (the owner was his youngest, most beloved son) as the sole manufacturer of the national car, named the Timor. The company, in a joint venture with KIA Motor from Korea, then started building factories to manufacture the Timor in Indonesia, but meanwhile the cars were being produced wholly in Korea and exported as CBUs to Indonesia without the paying of any import duty. The claim was that the Timor was a 'national' car. The car also enjoyed special fiscal treatment, such as reduced rates of value added tax. The decree imposed that the company should reach a local content requirement of 20 per cent in the first year, 40 per cent in the second year and 60 per cent in the third year, which was in accordance with the import duty incentive system ruled by the government through the 1993 regulation. This measure subsequently led to protests and the case was brought to the Dispute Settlement Body of the World Trade Organization (WTO) by Japan, the European Union and the United States. In the end, Indonesia lost the case and was forced to cancel the special status of Timor Putra Nasional. Subsequently, the WTO imposed sanctions on Indonesia to reduce the import tariff rates on automobiles and forced the government to open imports of CBU cars in June 1999, abolishing the incentive system based on local content achievement (see also Gunawan 2001: 2–3). The June 1999 regulation, set import duty for all categories of automobiles in CKD condition at 25 per cent, except for sedans, for which it ranged from 35 to 50 per cent, depending on the engine capacity. The higher the engine capacity, the higher the corresponding import tariff rate. Import duty for CBU commercial cars was fixed at 40 per cent, and for sedans it ranged between 60 to 75 per cent. On top of that, the government imposed a luxury tax on automobiles, whether wholly imported or assembled locally. The latest stipulation, dated May 2002, applies a tax rate between 10 and 75 per cent on commercial vehicles and between 30 to 75 per cent on sedans.

Sadly, since the new reformist but short-lived regime under President Abdurrachman Wahid from October 1999 to July 2001, there has been practically no industrial policy in the automotive sector. Though import tariff rates for CBU and CKD cars are still high, there are no restrictions on imports any more. The result is that, presently, all kinds of car brands can be imported, including very expensive cars like the Jaguar, and expensive models of leading brands such as Mercedez-Benz, BMW, Land-Rover and Toyota Landcruiser (Gaikindo 2001: 38–40).[1]

On the question of technology transfer

Basically, there is no difference between car producers around the world concerning technology transfer to developing countries. Technological inventions need a long process of R&D involving highly qualified researchers and scientists before they can be accepted for mass production. It is a process of long experience and accumulation of knowledge, and involves many failures, trials, testing, improvements, sophistication, a large amount of capital and, finally, successes. Hence all these inventions are very costly. No producer would transfer its technological knowledge to others, let alone in any free and voluntary manner. On the contrary, a producer will protect its inventions by all possible means and keep them secret, because they concern the existence of the producer's own enterprise and command over the market. Therefore, it is understandable if car producers are reluctant to transfer their technologies without compensation to their agents in developing countries, especially as they might end up becoming competitors in the future.

On the other hand, developing countries must fight hard to obtain new and advanced technologies. Generally, scientists and experts from advanced countries are quite willing to share their technological knowledge and experience with technicians from developing countries, if the people from developing countries are active in questioning and do not wait passively to be lectured. The easiest way to obtain advanced technology is, of course, through stealing, which is illegal and not ethical, despite the fact that it is practised worldwide. There are also, however, old and obsolete technologies that have become common goods and can be used freely.

Since technology development involves the costs of research and developing inventions, it has a price and can therefore be purchased on the market. There are patents, licences, royalties, fees and many kinds of transfer arrangements. Thus the question is not how far the owner of a technology is willing to transfer the technology, but at what price the owner is ready to transfer it and whether the buyer is willing to pay the price. Jusmaliani and Ruky found that the majority of around 150 automotive component producers had a technology agreement with overseas principals, largely following the customer–supplier relationships of the overseas principal and its suppliers. Technical assistance agreements involved a royalty fee between 1 and 3 per cent of the selling price, which constituted a prerequisite for exporting to Japan. Hence overseas principals play a major role in the development of component manufacturers (Jusmaliani and Ruky 1993: 47). Japanese investors also play an important role in the manufacturing of automotive components in Indonesia. A picture of Japanese involvement in Indonesia can be seen in the number of approvals of foreign investments in 1995 (see Table 6.4) and the structure of Japanese investments (see Table 6.5).

On the other hand, there are limitations in the ability of developing

Table 6.4 Approvals of foreign investment in the automotive component industry in Indonesia in 1995

Country	No. of projects	Investment (US$m)
Japan	10	116.0
Singapore	2	24.9
USA	1	3.0
Taiwan	1	5.6
Sweden	1	2.8
Joint ventures	1	11.0
Total	16	163.3

Source: Tarmidi (1996: 19)

Table 6.5 Japanese foreign investment in the automotive component industry in Indonesia

No.	Company name	Share owner	Equity (%)	Year	Capital	Workers	Products
1	Daihatsu Indonesia	Daihatsu Motor Nichimen Corp. Astra Int'l	20 10 70	1979	US$8m	230	Components for commercial vehicles
2	Daihatsu Engine Mfg. Indonesia	Daihatsu Motor Nichimen Corp. Astra Int'l	30 10 60	1985	US$3m	115	Automobile engines
3	Imora Honda	Honda Motor Kanematsu Imora Motor	45 15 40	1978	—	132	Chairs, exhaust pipes Tanks
4	Honda Prospect Engine Mfg.	Honda Motor Kanematsu Imora Motor	55 15 30	1986	IDR 7,544m	56	Automobile engines
5	Hino Indonesia Mfg.	Hino Motor Sumitomo Corp. National Motor, etc.	30 30 40	1982	US$5m	10	Diesel engines, pressed parts
6	Mesin Isuzu Indonesia	Isuzu Motors C. Itoh & Co. Toyo Menka Gaya Motor Garmak Pantya Mota	25 8 8 10 20 20	1985	IDR 4,432m	95	Diesel engines Components
7	Mitsubishi Krama Yudha Motors & Mfg.	Mitsubishi Motors Mitsubishi Corp. Krama Yudha, KTB	25.4 25.4 49.2	1975 1988	IDR 6,377m	634	Sheet metal parts Automobile engines
8	Suzuki Indonesia Mfg.	Suzuki Motor Indokarmo Utama	55 45	1976	US$ 7,400m	494	Automobile and motorcycle engines
9	Suzuki Engine Industry	Suzuki Motor Musyawara	49 51	1976	US$6m	150	Automobile and motorcycle engines
10	Toyota-Mobilindo	Toyota Motor Astra Int'l	56.7 43.3	1976	IDR 3,735m	878	Automotive components
11	Toyota Engine Indonesia	Toyota Motor Astra Int'l	51 49	1984	IDR 4,389m	62	Automobile engines

Source: Dodwell Marketing Consultants (1990: 663–664)

countries to absorb modern advanced technologies. First, the price for purchasing new technologies might be too high. The new technology might need large investments, and firms in developing countries are generally short of capital. The big question for technology adoption is whether the human resources in the receiving country are qualified and ready for absorption. The last constraint to investing in new technologies is the market size, given that local automotive component manufacturers are generally not allowed to export their products (compare Jusmaliani and Ruky 1993: 47). For the present, Indonesia can produce a maximum of 400,000 cars a year, a total that comprises many brands, types and models. In addition, there is the replacement market. The demand for a certain automobile component part is therefore not large enough to justify efficient and low-cost mass production, unless part of the production can find an export market.

However, it would be difficult for Indonesia, as a latecomer, to enter the world market of automobile components; there are already a large number of suppliers in the market, and competition is very intense for a newcomer – though this is no excuse for not entering the world market: some Indonesian component manufacturers have proved successful, albeit in some cases through joint ventures with foreign firms. Indonesia exports some automobiles in CBU and CKD condition, but their value is relatively small. Indonesia is rather more successful in exporting automobile components, and their value increased from US$344 million in 1996 to US$513 million in 2000 (see Table 6.6). Exports of motorcycle components by GIAMM (the Indonesian Automotive Parts and Components Industries Association) members, though still relatively small, increased from US$12.5 million in 2000 to US$15.2 million in 2001 (see Table 6.7). The achievements in exporting automobile and motorcycle components do not point to success in technology transfer, however, because most of the firms are foreign owned, be it fully or partially.

Hence the question of technological mastery is not a matter of transfer of technology per se, but a question of what the Indonesian government has done to promote the development of technology through education, training, promotion of research activities and legal protection of intellectual property rights. Though Indonesia has laws on patents, trademark, intellectual property rights, industrial design, trade secrets and layout design of

Table 6.6 Automotive industry exports from 1996 to 2001[a] (US$ million)

Year	CBU	CKD	Components
1996	33.7	1.6	343.9
1997	33.8	3.8	282.8
1998	34.3	6.5	376.3
1999	54.0	10.6	457.2
2000	64.7	20.4	513.3
2001[a]	23.6	0.7	315.7

Source: Ministry of Industry and Trade, 2001

Note
a January to October

Table 6.7 Exports of motorcycle component products by GIAMM members, 2000–2001

Component product (motorcycle)	Units		Type of units	Value (US$1,000)		Export destination
	2000	2001		2000	2001	
Battery	1,178,000	1,014,000	pcs	4,040	3,542	Europe, Africa, South Africa, Asia
Brake system	400,773	513,514	sets	2,958	4,782	Thailand, Japan
Cam chain	82,800	65,900	pcs	48	38	Hong Kong
Drive chain	31,200	31,760	pcs	63	73	Vietnam, Malaysia
Engine valve	1,860,419	1,610,200	pcs	1,621	1,484	Japan
Front fork and oil unit cushion	120,000	201,000	pcs	2,095	3,446	Brazil, Colombia
Shock absorber	248,000	213,000	pcs	1,697	1,768	Philippines, Spain
Speedometer	—	1,327	units	—	19	Thailand
Total				12,522	15,152	

Source: GIAMM

integrated circuits, the existing laws still need substantial improvements, and law enforcement is weak. There is a lack of research in the country and the number of patent listings is still small. Generally, firm owners are already satisfied with their businesses and look upon R&D as wasteful expenditure. Large firms usually have a R&D department, but it is not equipped with many or highly qualified personnel, and little money is spent on it. In addition, the availability of a well-developed industrial base, as a prerequisite that can support the development of downstream industries as in the Asian newly industrializing economies (NIEs), is also lacking.

Japanese transplants and technology transfer

The Indonesian government often complained that Japanese automobile producers, or principals, were very slow in transferring their technology to their Indonesian affiliates. Since the 1970s, and despite strong pressure from the government, the Indonesian automotive industry has still not been able to manufacture most components locally. What progress there was involved the components being manufactured by Japanese subsidiaries or by joint ventures. Critics in Indonesia often blamed the Indonesian firm-owners because they were, in the first place, merchants whose objective was to make quick, high profits and they were not dynamic industrialists by nature. However, the criticism was based on superficial observations because, meanwhile, the so-called merchants had invested large amounts of capital to build factories and invest in heavy machines. They were no longer simply assembling, but had gained substantial technological expertise and employed a relatively large pool of workers.

The failure to make much progress in technology transfer in the Indonesian automotive industry was largely due to the government's failure to have a firm policy directive. The government was indecisive and drifted between various interests of firm-owners, principals and government officials (Furqonny 1997).

A look at the extent of technology transfer in the automobile assembly industry also requires a look at the development of automobile component production and exports in the country. The higher the local content proportion in the final product of a car, the more successful is the technology transfer from abroad. In developing countries such as Indonesia, practically no development or inventions in automotive technologies have taken place, and they depend entirely on foreign car producers. Therefore, the only way for developing countries to develop an automobile industry is through technology transfer.

The large number of automotive component manufacturers and the capacity to export are a clear indication that automotive components can already be manufactured locally and that some can even compete on the world market. There are some 155 firms producing various automotive components in Indonesia (see Table 6.8).

Although Indonesia is already able to manufacture a relatively large number of automotive component items, production is still relatively small

Table 6.8 Number of firms in the Indonesian automotive component industry in 2002

Item	No. of firms	Item	No. of firms
Engine	9	Brake lining	4
Filter	8	Brake shoe	5
Piston	4	Seat and seat frame	8
Transmission	5	Safety belt	4
Drive axle	6	Control cable	2
Clutch	4	Door lock	3
Brake system	4	Evaporator	1
Cylinder block	3	Exhaust manifold	3
Chassis and body	9	Floor mat	1
Shock absorber	5	Gasket	4
Fuel tank	9	Horn	3
Radiator	3	Intake manifold	2
Spark plug	3	Coil spring	2
Spring	5	Muffler	4
Wheel rim	10	Rubber parts	4
Wiring harness	5	Safety glass	3
Battery	8	Interior	2
Alternator	3	Alternator and starter	3

Source: Ministry of Industry and Trade, 2002

(see Table 6.9). However, this does not necessarily mean that technology transfer has taken place, because most manufacturers are subsidiaries of their foreign parent companies and are fully foreign owned. Therefore, it is doubtful whether one can speak about technology transfer in this case, because only the site of production has been relocated to Indonesia. Thus to increase the production of automotive components locally, more foreign direct investments should be invited into the country. Only in cases of joint ventures or technology transfer agreements can one speak about technology transfer.

The main obstacle to technology transfer is that the principal–sole agent relationship is very restrictive. The sole agent, the automobile assembler in Indonesia, is not free to import any type or model from the principal, because the principal decides which types of cars can be exported in CKD condition to Indonesia (Jusmaliani and Ruky 1993: 56).

Owing to the mandatory deletion programme, the import content in the automobile assembly industry has fallen, but the import content proportion in the automotive component industry has risen, because the automotive component industry, in turn, is highly dependent on imports. This is because specific material must still be imported, owing to quality requirements and to domestic unavailability. In the motor vehicle component industry, the portion of imported raw material was 49 per cent of gross output in 1995, and in the motor vehicle assembly industry, 13 per cent. The more items are domestically produced, the higher the share of the imported material rises (Sato 2001: 3–14, 15, 16).

According to a survey conducted by the author, the main constraints to technology transfer in the automotive component industry were a lack of

Table 6.9 Production of automobile components in 1989/90 and 1995/96–1998/99[a]

Component	Unit	1989/90	1995/96	1996/97	1997/98	1998/99[a]
Shock absorber	000 unit	1,202	1,729	1,816	1,903	313
Radiator	000 unit	171	435	543	575	419
Exhaust system	000 unit	312	1,394	1,533	1,176	193
Filter element	mill. set	3.6	5.5	12	14	13
Piston	000 unit	570	1,337	1,471	1,703	280
Piston ring	000 unit	3,010	4,758	5,645	2,103	345
Sparkplugs	mill. unit	27	41	34	23	25
Diesel engine	000 unit	36	76	85	108	81
Gasoline engine	000 unit	157	216	235	220	165
Cabin	000 unit	128	178	187	376	64
Chassis	000 unit	183	302	317	324	64
Axle	000 unit	138	135	143	184	53
Propeller shaft	000 unit	138	139	145	145	1
Rear body	000 unit	53	107	112	37	64
Brake system	000 unit	273	410	430	376	64
Wheel rim	000 unit	760	1,933	2,043	2,308	917
Fuel tank	000 unit	144	485	470	376	64
Leaf spring	000 ton	22	51	38	22	23
Seat and seat frame	000 set	244	741	778	376	192
Clutch system	000 set	130	546	573	376	64
Transmission	000 set	147	294	309	329	53
Steering system	000 set	134	274	287	329	53

Sources: Republic of Indonesia (1995b: XI/38; 1997: XI/37; 1999: 42)

Note
a Temporary figure

technological expertise, an insufficient number of skilled workers and the unavailability of material. The availability of capital, on the other hand, did not seem to be a major problem (Tarmidi 1998: 74). Quality of human resources in developing countries primarily depends on good education, and this is widely lacking in Indonesia. Indonesian human resources are ready to adopt low- and mid-level technologies, but are considered not ready yet for high-level, sophisticated technologies.

More important than technology transfer is the transfer of expertise from industrialized countries to developing countries such as Indonesia. This can be done in two ways: first, by sending Japanese skilled workers to supervise Indonesian workers in Indonesia, particularly in Japanese subsidiaries or joint ventures (this is already being done); second, by sending Indonesian workers to factories in Japan for OJT to gain experience and knowledge. Both ways were already in effect in Japanese firms in Indonesia (Thee 1994: 53). However, the number of workers sent to Japan was still limited. In the case of the Timor automobile, hundreds of young workers were being trained in KIA factories in Korea. The above are positive measures towards increasing industrial expertise among Indonesian workers.

Another way of looking at technology transfer is by assessing technical assistance programmes provided to Indonesia by the government of Japan. Japan International Cooperation Assistance (JICA) had various technical assistance projects in Indonesia; however, only a few involved some form of transfer of industrial technology and skills. There is also a big constraint in that Japanese experts tend to be weak in foreign languages (Thee 1994: 49, 50).

The extent of technology transfer also depends on the price class of certain car brands being assembled in Indonesia, such as Volvo, Mercedes-Benz and BMW. Unlike brands with a relatively large production volume, these expensive cars demand very high-quality component products and their sales are relatively small; hence it is difficult to increase the local content of parts in the final product.

Unlike European or American firms, Japanese firms are often criticized for keeping managerial positions for Japanese nationals only and not appointing other nationals to these positions. The Japanese management style is quite different because it has a lifetime employment system, and promotion to top positions, based as it is on a seniority system, can take twenty years or more. In general, Japanese firms in their operations do not rely on detailed job descriptions, as is the case in Western firms, but rather on long working experience and OJT (Thee 1994: 51, 52, 55).

As of October 2002, there are 121 automotive component manufacturers listed as members of GIAMM, of which sixty-eight firms are domestic companies, forty-six are joint ventures with Japanese investors, two are joint ventures with Germany, another two with the United States, two are joint ventures with Taiwan and Korea respectively and one is a state-owned enterprise. In addition, there are seven manufacturers that are inactive members and some others that are non-members of GIAMM, but the number is unknown. The Astra Auto Component Group alone has thirty-two subsidiary firms. It is interesting to look at the Astra investment structure: there are five domestic firms, thirteen Japanese foreign direct investments, thirteen domestic firms with Japanese technical assistance and one domestic firm with technical assistance from Sweden.

Conclusion

Indonesia started assembly activities in the automobile sector in 1964 and moved to assembling completely knocked down kits in 1971. Since the early 1980s, Indonesian governments have tried various policy measures to make the sector more competitive and to increase the portion of locally made components. To meet this second objective, compulsory deletion programmes were in place between 1979 and 1993. After 1993, the deletion programme was substituted by a tariff incentive system. These policies have resulted both in higher percentages of Indonesian components in cars produced for the local market and in higher exports of parts and components. However, these components are very often produced by Japanese companies in Indonesia, by Japanese–Indonesian joint ventures, or by Indonesian companies operating under technical agreements with Japanese principals. Technology transfer from Japanese actors to Indonesian actors takes place but is hampered by a lack of technical expertise, undeveloped local support industries and by there being an insufficient number of skilled workers and engineers.

Note

1 In Indonesia, these luxury cars cost between Rp. 1 billion and Rp. 6 billion, or roughly between US$110,000 and US$660,000.

References and further reading

'Automotive Industry Enters a New Phase', *Data Consult/ICN*, 192 (25 March): 7–22.

CIC (1996) 'Developments of Automobile Industry and Market in Indonesia', *Indocommercial*, 150 (4 April): 3–28.

Dodwell Marketing Consultants (1990) *The Structure of the Japanese Auto Parts Industry*, 4th edn, Tokyo: Dodwell Marketing Consultants.

Furqonny, R. (1997) 'Proses Alih Teknologi Otomotif' (The Process of Automotive Technology Transfer), *Kompas*, 7 August: 20.

Gaikindo (Association of Indonesian Automotive Industries) (2001) *GAIKINDO Profile 2001*, Jakarta: Gaikindo.

Gunawan, R. (2001) 'The Short Analysis of Motorcycle's Market and Industries in Indonesia for the Year 2000 and First Half 2001', paper presented in New Delhi, 5 September.

Hansen, J.R. (1971) 'The Motor Vehicle Industry', *Bulletin of Indonesian Economic Studies*, 7 (2) (July): 38–69.

Hill, H. (1995) 'Indonesia's Great Leap Forward? Technology Development and Policy Issues', *Bulletin of Indonesian Economic Studies*, 31 (2) (August): 83–123.

Jusmaliani, and Ruky, I.S. (1993) *Transfer of Technology from the Perspective of Host Countries*, Jakarta: Indonesian Institute of Sciences.

Republic of Indonesia (1993) Decree no. 114/M/SK/6, Jakarta: Minister of Industry.

—— (1995a) Decree no. 108/M/SK/5, Jakarta: Minister of Industry.

—— (1995b) *Realization of the 1st Year of the 6th Five-Year Development Plan, Addendum to the Presidential State Address before the Parliament* (16 August), Jakarta: Ministry of Information.

—— (1997) *Realization of the 3rd Year of the 6th Five-Year Development Plan, Addendum to the Presidential State Address before the Parliament* (16 August), Jakarta: Ministry of Information.

—— (1999) *The Realization of Reformation in Development, Addendum to the Presidential State Address before the Parliament* (16 August) Jakarta: Ministry of Information.

Sato, Y. (1998) 'The Machinery Component Industry in Indonesia: Emerging Subcontracting Networks', in Sato, Y. (ed.) *Changing Industrial Structures and Business Strategies in Indonesia*, Tokyo: Institute of Developing Economies.

—— (2001) 'Structure, Features and Determinants of Vertical Inter-firm Linkages in Indonesia', unpublished Ph.D. dissertation, University of Indonesia, Jakarta.

Shauki, A. (1996) 'The Indonesian Automotive Industry', paper presented at the ASEAN Automobile Workshop, Bangkok, 29–30 March (unpublished).

Sudarsono, E. (1996) 'Direktur Jendral Industri Logam, Mesin dan Kimia' (Indonesia's Perspective on the Automotive Industry), paper presented at the international symposium 'The Asian Age of Automotive Industry', Jakarta, 17 December (unpublished).

Surajaya, I.K. (1983) 'Alih Teknologi Jepang ke Negara-negara ASEAN' (Transfer of Japanese Technology to ASEAN), *Kompas*, 3 October.

Tarmidi, L.T. (1996) 'Japan's Role in the Development of Support Industries in Indonesia', paper presented at the international conference 'Japan's Role in the Development of Support Industry to ASEAN Nations', Institute of East Asian Studies, Thammasat University, Bangkok, 9–10 May (unpublished).

—— (1998) 'Industri Komponen Otomotif' (The Automotive Component Industry), in Perhimpunan Alumni Jerman (Union of German Alumni), *Daya Saing Industri Indonesia* (The Competitiveness of Indonesian Industries), Jakarta: Perhimpunan Alumni Jerman.

Thee, K.W. (1994) 'Technology Transfer from Japan to Indonesia', paper presented at the conference 'The Transfer of Science and Technology between Europe and Asia, 1780–1880', organized by the International Research Center for Japanese Studies, Kyoto, Japan (unpublished).

Thee, K.W. and Pangestu, M. (1994) 'Technological Capabilities and Indonesia's Manufactured Exports', Jakarta (unpublished paper).

7 Strategic effects of firm sizes and dynamic capabilities on overseas operations

A case-based comparison between Toyota and Mitsubishi in Thailand and Australia[1]

Takahiro Fujimoto and Shinya Orihashi

The purpose of this chapter is to analyse how the behaviour of multinational enterprises' (MNEs') overseas operations, in their efforts to adapt to local environmental changes, is affected by such firm-specific characteristics as firm size and organizational capability. Our empirical case study focuses on the overseas production operations of two Japanese automobile MNEs, Toyota Motor Corporation (Toyota) and Mitsubishi Motors Corporation (Mitsubishi Motors) in both Thailand and Australia.

In international business, much attention has been directed at the international expansion of MNEs based on competitive advantages they created in their home countries that enable them to compete effectively against host-country firms (Vernon 1971; Hymer 1976; Dunning 1980; Bartlett and Ghoshal 1989). More recently, the organizational capabilities and competitive advantages of Japanese manufacturing firms (in automobiles, electronics, etc.) have been analysed as important factors in the establishment of overseas transplants (cf. Abo 1994).

This approach has become a standard theoretical framework for analysing the fundamental issues of international business, such as the reasons why MNEs exist in the first place. However, this model, which emphasizes the application of country-specific resources, is often insufficient for explaining why firms from the same home country pursue different development paths and employ different strategic choices when managing their overseas operations.

First, we need an additional framework to explain why the competitive behaviour of MNEs is so firm specific. The existing literature tends to emphasize country-specific behavioural patterns and performance of MNEs in general (e.g. advanced versus developing countries, Japanese versus Western firms), while it de-emphasizes the inter-firm differences between MNEs from the same country. However, significant differences in the local operational patterns of, for example, the two Japanese multinational automobile manufacturers considered in this chapter are easily observable. For this reason, we need to turn our attention to firm-specific factors, such as firm size and organizational capabilities.[2]

Second, we need to stress the dynamic aspects of MNEs' organizational capabilities. Explaining away the competitive actions of a local subsidiary as simply an application of resources developed in the home country tends to dismiss local actions as passive responses to the local environment and to the policies of headquarters. The capability-building processes of the operations overseas have themselves received little attention. However, when confronted with a local crisis, local operations have been known to play an active role in improving their organizational capabilities in response. In place of a linear and static framework to explain this phenomenon, what is needed is an analysis that examines the 'dynamic capability of capability building' that enables the MNE to build the capabilities in its local operation needed to respond to the crisis.

Third, as has been shown by Fujimoto (1998b) and Fujimoto and Sugiyama (2000), the capability development paths of foreign operations are emergent in nature. In other words, they do not necessarily proceed according to a deliberate plan, but are frequently rather processes fraught with unintended successes and failures, as well as trial and error. Therefore, a firm's 'evolutionary capability' (Fujimoto 1998a, 1999, 2000) to gain comparative advantage over rivals given a capability-building process so emergent in nature requires examination. That is to say, the behaviours of MNEs may be explained not only by their deliberate strategies, but also by their 'emergent global strategies'.

In summary, this chapter emphasizes three additional aspects of MNE behaviour in its local operations: firm specificity, dynamics of capability building and emergent processes.

Framework: internationalization of 'large' and 'small' MNEs

Firm size and organizational capability

Generally speaking, there are at least two main factors affecting firm behaviour and competitiveness in the same industry or market: firm size and organizational capability. The former is a factor emphasized by standard economics. Organizational capability, on the other hand, means a system of organizational routines that create firm-specific and difficult-to-imitate advantages. A firm's organizational capability consists of (1) static capability to consistently outperform rivals at any given point in time, and (2) dynamic capability that enables the firm to improve its performance and capability faster than rivals (Penrose 1959; Nelson and Winter 1982; Teece *et al.* 1997; Fujimoto 1999; Dosi *et al.* 2000). Thus when we observe differences both in patterns of performance and behaviour and in responses to environmental changes in two MNEs operating in the same local environment, we can infer that they may be different in their firm sizes, static organizational capabilities for higher productivity and quality, or dynamic capabilities for the improvement of their productive performance and capability.

Basic types of MNEs

On the basis of the above discussion, we propose a basic classification of MNEs that may provide additional insight for understanding differences in the local behaviour of MNEs (Table 7.1). This matrix classifies manufacturing MNEs according to (1) the relative size of the firm as a whole, and (2) the organizational capability of the firm in manufacturing operations (e.g. productivity, and quality in production and product development) at a certain point in time. Note that the following classification is basically static in that it disregards the dynamic capability of each MNE. This is because, at this point, we focus only on the initial conditions of the MNEs prior to a local crisis. Dynamic capabilities, with which the MNEs deal with a crisis, are discussed later on.

The resulting four categories are: competent large firms; competent small firms; incompetent large firms; and incompetent small firms. Note that 'competent' and 'incompetent' do not refer to the financial and managerial capabilities of an MNE as a whole, but to its operational capabilities that bring about higher productive performance in its factories and development centres. Note also that 'large', 'small', 'competent' and 'incompetent' are conceived of in relative terms. 'Competent firm' means that the firm's operational performance is ranked among the best worldwide; 'large firm' means that the firm is listed among the largest companies worldwide in the industry. According to this classification, Toyota and Mitsubishi Motors may be classified as a 'large competent firm' and a 'small competent firm', respectively.

It is also known that Toyota possesses a superior dynamic organizational capability, in particular an 'evolutionary learning capability' (Fujimoto 1998a, 1999, 2000). That is, Toyota may be characterized as a 'large competent firm that also learns and evolves rapidly'.

Stages of environmental changes

Let us turn to the dynamic processes on the environmental side. For the present analysis, we identify three stages: before and after the period of growth, and the crisis (Table 7.2). Although our empirical foci in this chapter are the Thai and Australian automobile markets during the 1980s and 1990s, a similar pattern may be observed in other emerging markets. In any case, for the present empirical purpose of analysing Japanese MNEs

Table 7.1 Analytical framework

Firm size	Operational capability	
	Relatively high	Relatively low
Relatively large	Large competent firm (dynamic capability) = high/low	Large incompetent firm
Relatively small	Small competent firm (dynamic capability) = high/low	Small incompetent firm

Table 7.2 Analytical framework (environmental side)

Initially	Production expansion stage	Production shrinkage stage
Conditions		
Local market is limited	Stronger local market or government's policy stimulates firms to increase production volume	Local market gets weaker, owing to an economic crisis or an increase in car imports
Importing is very difficult, owing to government policy		
Small company		
Local sales network is relatively weak	Local sales are still not sufficient for competitive volume, so they concentrate worldwide production of a specific model to the plant	Already started to export to various countries
Import-substituting plant	The mission of the plant changes, but new investment is limited, owing to the firm's financial constraints	Already achieved the capability to compete in many types of markets worldwide
Not exporting on a large scale	Must begin exporting on a large scale to various countries Capability building starts	
Large company		
Local sales network is strong	Increased local sales force the construction of a new, locally oriented plant	Now they have to begin exporting on a large scale to maintain their operation
Import-substituting plant	Still not exporting on a large scale	Capability building starts
Not exporting on a large scale		The main mission of the plant is still locally oriented; the scope of exports is limited, owing to the international division of production

facing a local crisis, the following three-stage framework will be capable of summarizing the local situation reasonably well.

Initial condition: small local market and import substitution policy

Less-developed car-producing countries in the 1960s and 1970s were generally characterized by small markets (less than 200,000 units per year) and trade policies by which national governments aimed at import substitution by restricting or banning complete vehicle imports. As a result, many of the automobile MNEs chose to establish knocked down (KD) assembly plants in each fragmented local market and produced small numbers of many different models, each of which was designed in the MNE's home country. Local governments also enforced local content policies that facilitated local production of a certain percentage of automobile parts.

The local facilities for assembly and parts production naturally lacked international competitiveness. Their exports were limited to low volumes headed for small neighbouring markets. In addition, the bandwagon effect[3] among the MNEs resulted in many automakers rushing to build assembly and parts plants in each of the small markets, creating an extremely fragmented production structure, with several thousands of vehicles per year per model being produced at most. As a result, the sizes of the local production facilities did not reflect the firm size of the MNEs in their home countries – the local production facilities were all small regardless of the MNEs.

Production expansion stage

The second stage is production expansion. The automobile MNEs, which had struggled with the chronic problem of small market size and inefficient small volume production in the earlier stage, were able to succeed in expanding local production to a certain extent.

In some cases, production volume expanded mostly as a result of the expansion of the local market itself. In many Asian and Latin American countries, for instance, increases in income per capita and the emergence of a middle class resulted in the expansion of both vehicles in use and annual sales volume in these local markets. This was particularly the case in South-East Asia from the late 1980s to the early 1990s. The Australian market, on the other hand, had been basically saturated for many years, but it still experienced some expansion thanks to a favourable business cycle and the creation of new segments (e.g. the emergence of the sport utility vehicle segment).

In other cases, exports contributed to the growth of local production. This was particularly the case when the host country's government adopted export promotion policies when the local production facilities had certain competitive advantages vis-à-vis neighbouring countries, or when there were some free trade markets near the country.

Whether the growth of the local facilities relied on local sales expansion or export growth may also depend upon the global strategies, financial resources, and competitiveness of the MNEs in question. As is discussed later, smaller MNEs may have a higher tendency to choose local production expansion through greater exports, partly because the number of their local facilities worldwide is comparatively limited (i.e. one local production facility is more likely to be shared by multiple markets).

In any case, responding to the expansion phase, the automobile MNEs sometimes replaced their ageing KD assembly plants of earlier days with new and expanded plants. Such cases were observed in both Thailand and Australia during the 1990s, particularly in the case of large and cash-rich firms, such as Toyota.

Production shrinkage stage

The production expansion stage may be followed by a period of 'crisis for the local factories', the third stage, which is characterized by an unanticipated and sudden shrinkage of the production volume of the local facilities. Such crises may come about in at least three ways. First, a local economic crisis, such as the Asian currency crisis of 1997, may cause a collapse of local automobile sales. Second, a sudden change in government policy towards trade liberalization may cause a crisis for the local factories as a flood of imports crowd out inefficient local production. The Button Plan of the Australian government, implemented since the mid-1980s, is a typical example of this, and developing countries' participation in the World Trade Organization (WTO) after 2000 could be another. Third, a market collapse or protectionist measures by the main export destinations of the local factories could create a crisis for the local plant.

Dynamic capability for emergent global strategy

So far, we have proposed a simple model of four types of MNEs facing three stages of local environmental changes. It is obvious, however, that a firm with a strong dynamic capability is likely to adapt its strategy to the environmental changes more speedily and effectively than one without such a strong dynamic capability. The notions of 'emergent strategy' and 'evolutionary learning capability' may provide additional insights when analysing the dynamic capability of MNEs.

In the field of strategic management, the notion of 'strategy as plan', in which strategic intent precedes strategic implementation, has been a prevalent idea for many years (see, for example, Andrews 1980; Hofer and Schendel 1978). However, there has also been another concept of strategy, namely 'strategy as pattern', which assumes the possibility that competitive strategy may be formed even without a competitively rational prior intention. Mintzberg and his colleagues call a strategy that was unintended but realized 'emergent strategy' (Mintzberg and Waters 1985).

Actual strategy formation tends to be an unpredictable mixture of both emergent and deliberate strategies (Fujimoto 1998a, 1999, 2000). A firm's distinctive dynamic capability to create effective organizational routines in this kind of emergent situation may be called its 'evolutionary learning capability'.

Such an argument on emergent strategy and evolutionary capability may also be applied to the case of global strategy. In this chapter, we will also pay attention to 'emergent global strategy' (Fujimoto 1998b), in which a firm builds the organizational capabilities of its local facilities through *ex-post* responses to local crises. Further on in this chapter, we will focus on a company that has been shown to possess a distinctive evolutionary capability: Toyota Motor Corporation (Fujimoto 1999).

Predictions: responses to local crises by 'large' and 'small' firms

Assumptions

Basing our ideas on the above framework that emphasizes inter-firm differences in organizational capabilities, we now derive a few hypotheses, or predictions, that may help explain the behaviour of Japanese automobile MNEs facing crises in the local market. Assuming the stylized fact that Japanese automobile manufacturers in general tended to possess a high manufacturing capability during the 1980s and 1990s (Womack *et al.* 1990), the following hypotheses focus on the behaviours of competent firms. In other words, our hypotheses for the following empirical analyses deal with the competitive behaviour of competent firms, large or small, facing the same local market before a local crisis (i.e. the production expansion stage) and after the crisis occurs (i.e. the production shrinkage stage).

In the case analysis presented later in this chapter, Toyota can be regarded as a typical 'large competent firm', whereas Mitsubishi Motors is a typical example of a 'small competent firm'. This is because they are significantly different in size, but they are both competitive in production and development productivity.

Let us also assume that both the large competent firm and the small competent firm make use of their operational advantages in manufacturing and deploy overseas production facilities, and that their overseas facilities are located in the same set of host countries as a result of the oligopolistic bandwagon effect of MNEs. Also, the firms may possess a high level of dynamic capability (e.g. the evolutionary learning capability of Toyota; see Fujimoto 1999) in addition to their static manufacturing capability. On the basis of the above assumptions, we may derive at least three hypotheses of inter-firm differences in the local behaviour of the firms, as follows.

Prediction concerning the pre-crisis (production expansion) period

A large firm with greater financial resources will tend to set up a larger number of local-market-oriented production lines and develop a larger number of locally dedicated models than a small firm. If a large firm's model turns out to be close to a world model that can be sold in multiple markets, such a model may be produced in more than one factory internationally.

A small firm, on the other hand, needs to cope with a large number of overseas markets with a smaller number of production lines and models. Therefore, it tends to be more oriented to a world model that is shared by multiple markets, as well as export-oriented production lines aiming at global markets, compared to the large firm. This tendency applies to both home-country and overseas factories. If customers in each local market have unique needs that may be filled only by a country-specific model, or if the production control cost for handling multiple export models is high, then, all other things being equal, the small firm may suffer from a competitive disadvantage vis-à-vis larger firms.

Thus we predict that a large competent firm, with all other things being equal, has a greater tendency to have local-market-oriented production lines and locally dedicated models than does a small competent firm with similar operational capabilities.

Prediction concerning short-term effects in the post-crisis (production shrinkage) period

Let us now assume that a crisis for a local plant (e.g. a surge of imports, collapse of the local market) occurs, which dramatically decreases shipment volume from the local factory to the local market. In this situation, a small firm, which has had to rely on export-oriented production lines and product models during the expansion period (hypothesis 1), is more likely to enjoy unintended advantages vis-à-vis a large firm, because the former's locally produced models are less dependent on the local market, which is in crisis. In other words, the small firm has been forced to build manufacturing capabilities to cope with many overseas markets because of its smaller number of products and lines, and these turned out to absorb the local shocks better once the crisis happened. This phenomenon may be regarded as a global emergent strategy, as the small firms enjoy unintended but realized competitive advantages.

A large firm, in contrast, has product models and production lines dedicated to the local market. This was regarded as advantageous during the expansion period because they fitted more flexibly to local demands, but it turns out to be a disadvantage in the production shrinkage stage, because the production line cannot absorb the shock via export expansion. Thus, at least temporarily, the large firm is likely to suffer more from the local shock (Sugiyama and Fujimoto 1999).

Our second prediction, therefore, is that a large competent firm with a higher reliance on local-market-oriented production lines and locally dedicated models is affected more seriously by the local crisis of sharp demand decrease than a small competent firm, all other things being equal.

Prediction concerning local capability building in the post-crisis (production shrinkage) period

What, then, should firms facing the unexpected local crisis do? Small competent firms, with poorer financial resources at their headquarters, will try to minimize additional local investment and accelerate export drives, thereby absorbing the shocks more effectively. Since the small firms tended to rely more heavily on exports during the expansion period (hypotheses 1 and 2), further export expansion is likely to be easier for the smaller firms than for the larger firms.

A large firm, on the other hand, is more likely to suffer a bigger production slump because its plants tend to be more dependent on the shrinking domestic demand (hypotheses 1 and 2), owing to over-adaptation to the local market. If the company's dynamic capability is low, then it will try to survive by minimizing investments and simply waiting for the recovery of the market. In an extreme case, the company may close the local facility.

If the large firm has not only a high static manufacturing capability but also a high level of dynamic capability of capability building, the company may combine such a dynamic capability and the necessary financial investment for converting its local-market-oriented plants to export-oriented ones through rapid capability building for exports. The local subsidiaries may play a leading role in executing such a change in the local manufacturing capability. If the large firm succeeds in the export capability building of its local plants in many countries and enriches its global logistics network among its complementary production facilities, that MNE may change itself from a multi-domestic firm to a global firm (Porter 1985). This globalization, however, is not necessarily based on a deliberate global strategy, but on *ex-post* trial and error in response to the crisis. In this sense, it may be regarded as emergent global strategy (Fujimoto 1998b).

Our third prediction, then, is that when facing a local crisis, a large competent firm with a high dynamic capability of capability building tends to build its local plants' export capability in order to minimize the shock of the local demand shrinkage, all other things being equal.

Having illustrated our analytical framework and a few predictions derived from it, the next step is to examine how this framework can explain the actual competitive behaviours of manufacturing MNEs facing expansion and subsequent local crisis. As this chapter is exploratory in nature, aiming at theory building rather than rigorous prediction testing, a preliminary case analysis fits this purpose.

On the basis of our problem setting and analytical framework, the remainder of the chapter will investigate the cases of Toyota (a typical large competent firm with a high dynamic capability) and Mitsubishi Motors (a typical small competent firm) and their local operations in Thailand and Australia. In each case, we will examine whether the two firms responded to local expansion and subsequent local crisis in a manner consistent with our framework and predictions. We will pay special attention to the dynamic and emergent aspects of their local capability building.

Comparison between Toyota and Mitsubishi Motors

The factors we considered in selecting these firms for our study are that they manufacture and sell similar products, originate in the same country, and have manufacturing enterprises in the same host countries. The two Japan-based automotive manufacturers, Toyota and Mitsubishi Motors, meet these conditions in both Thailand and Australia. The principal characteristic Toyota and Mitsubishi Motors have in common is, in short, that they both operate internationally with their competitive advantages in manufacturing and R&D. First, both companies have already established a lean manufacturing system and they maintain competitive advantage internationally in both productivity and quality (Womack *et al.* 1990). In addition, they also maintain a competitive advantage in both overall product development performance and lead time of product development (Clark and Fujimoto 1991). Second, both companies have overseas operations in nearly the same locations, including Thailand and Australia, a situation that can be considered a result of the bandwagon effect.

As a result, both companies faced almost the same environmental changes in the Asian automotive market, namely its rapid growth in the early 1990s (specifically until 1996) and sudden shrinkage in the late 1990s. On the other hand, in Australia, where the automotive market has matured, the two companies were initially subject to the same import substitution policy, which aimed to protect the local market from international competition. They also later faced the same sudden shrinkage of the market for locally assembled automobiles. Moreover, both firms have engaged in KD operations for supplying the local market in both Thailand and Australia since the 1960s.

Despite the above-mentioned similarities, there are also significant differences between the two companies in terms of corporate scale and financial resources, both in Japan and worldwide. Toyota is the leader of the Japanese automobile market, where it manufactures more than 3.4 million vehicles per year domestically, and about 5 million worldwide. Toyota's assembly and component plants are mainly concentrated around Toyota City, Japan. Moreover, Toyota has capital tie-ups with Daihatsu Motors, which is strong in micro-cars, and Hino Motors, which is strong in large buses and trucks. Toyota has left the manufacture and sales of micro-cars, large buses and trucks to these manufacturers. Furthermore, Toyota has left the R&D and manufacturing of Toyota-brand light trucks, mini-vans and small buses to Kanto Auto Works and Toyota Auto Body, two of Toyota's subsidiaries, as well as Araco Corporation. Toyota itself has thus been able to focus its resources on the R&D and manufacture of passenger cars. We should therefore view Toyota's actual strength in R&D and manufacturing as going far beyond what can be found in Toyota proper.

Turning to Mitsubishi Motors, we see that it is the fourth largest firm in the Japanese automobile market, and it manufactures about 1 million vehicles domestically and about 1.8 million worldwide. Its main plants in Japan are located in Nagoya and Kurashiki, which are about 400 kilometres apart. Mitsubishi Motors has no subsidiaries to which it can leave its R&D or manufacturing functions, so in order to maintain its full line-up of vehicles, it has had to develop and manufacture vehicles ranging in size from micro-cars to large buses and trucks. Naturally, its resource allocation towards R&D and manufacturing has had to be somewhat broad and thinly spread. This restriction has affected the firm's decision making regarding overseas operations.

Throughout the 1990s, Toyota's sales were roughly triple those of Mitsubishi Motors in both domestic and worldwide production. Moreover, Toyota's operating income has been more than ten times that of Mitsubishi Motors. Naturally, these differences also produce differences in the financial resources available for product development and new models.

In order to operate widely overseas, much the same as in the case of Toyota, the foreign subsidiaries of Mitsubishi Motors, with its limited financial resources, have traditionally been joint ventures, and the firm has not been too particular about possessing a majority equity stake. Such firm behaviour contrasts remarkably with that of Toyota, which has sought to operate its foreign subsidiaries from a majority stake.

Toyota is noted to have a relatively high dynamic capability, or 'evolutionary learning capability', meaning that the firm is able to evolve its organizational routines, either planned or emergently, over the long term. The so-called Toyota Production System (TPS) can also be regarded as a result of this emergent capability. There is no academic research that has rigorously clarified that Mitsubishi Motors is somehow inferior in this respect, but there is much historical evidence to indicate that Toyota has a remarkable dynamic organizational capability (Fujimoto 1998a, 1999).

In summary, both Toyota and Mitsubishi Motors can be classified as having a high static manufacturing capability for productivity and production quality. Also, they are similar in terms of the geographical pattern of their overseas operations. However, there are great differences in company size between the firms. Moreover, Toyota may have a higher dynamic capability.

Case study 1: the Asian economic crisis and Toyota/ Mitsubishi Motors in Thailand (Table 7.3)

Overview of the automotive industry in Thailand and time period of the study

In Thailand, there has never been an industrial policy that excluded foreign control of local automobile manufacturers. Rather, the industrial policy focused on import substitution using foreign-designed vehicles assembled by foreign-owned local manufacturers, including having parts produced locally. So, foreign-owned manufacturers assembled their vehicles solely for the local market, and these firms included Japanese auto manufacturers. However, a major environmental change took place in the 1990s, and foreign manufacturers, including the two companies targeted in this research, had to change their strategy in Thailand and its vicinity. Within our framework, we consider these changes by dividing them into two stages (Table 7.4).

Production expansion stage

From the early 1990s to 1996, because of the country's rapid economic growth, automotive sales in Thailand grew rapidly and manufacturers were

Table 7.3 Toyota and Mitsubishi Motors in Thailand

	MMC Sittipol	Toyota Thailand
Started production in Thailand	1966	1964
Capital (in millions of bahts)	834	4,520
Shareholders		
Japanese	Mitsubishi Motors 46.2%	Toyota 69.6%
Thai	MHTC 52.0%, Lee Group 1.7%	TABT 15.5%, Siam Cement 10%, etc.
Employees (2000)	2,945	4,041
Japanese nationals (2000)	36	35
Thai directors (1999)	6	4
Sales (1999; in millions of bahts)	39,038	46,445
Annual production capacity	136,000	240,000
Pick-up exports to	90 countries other than North America (mainly South Europe and Australia; no export to North America)	Australia, Cambodia, the Philippines, Laos
Models		
Passenger car (sedan)	Lancer	Camry, Corolla, Soluna
Commercial car	Strada (pick-up), Canter (medium-sized truck), Fighter (king-size truck)	Hilux 4/2 (pick-up), Dyna (truck)

Source: JAMA, firm interviews in June 1999 and November 2000

obliged to increase their production. The automotive market in Thailand was forecast to increase to more than 1 million vehicles a year; so many manufacturers increased their total production capacity.

Production shrinkage stage

After 1997, because of the baht's depreciation in 1997, automotive sales in Thailand fell sharply. In 1998, they fell to about 150,000, about one-quarter of the 1996 total. So, it became necessary for manufacturers to start exporting, and rapidly increase their exports in order to survive.

From the 1960s, both Toyota Motor Thailand (TMT) and Mitsubishi Motors' local affiliates engaged in small-scale KD assembly, but were operating only import substitution plants under the Thai government's regulations on imported vehicles. Mitsubishi Motors' KD plant merged with its sales company in 1987, forming MMC Sittipol Co., Ltd (MSC).

Production expansion stage

MSC's main strategy in the production expansion stage was, in short, the construction of an export-oriented plant for 1-ton pick-up trucks (pick-ups).

Table 7.4 Analytical framework (environmental side) – Thailand

Initially	Production expansion stage (early 1990–1996)	Production shrinkage stage (mid 1997–)
Conditions		
Local market is limited	Stronger local market forces them to increase their production volume	Local market gets weaker, owing to Asian economic crisis
Importing is very difficult, owing to government policy		
Mitsubishi Motors		
Local sales power is relatively weak	A new plant is built in addition to the existing plant, with a concentration on worldwide production of one-ton pick-up trucks	Already started large-scale exporting to various countries
Import-substituting plant	The mission of the plant changes, but investment is limited by outsourcing plant operations to suppliers	Already achieved the capability to face many types of markets worldwide
Not exporting on a large scale	Starting to export on a large scale to various countries	Shuts old plant and concentrates its production in new plant
	Capability building starts	
Toyota		
Local sales power is strong	Increased local sales stimulate the construction of a new, locally oriented plant	Must begin large-scale exporting to maintain operation
Import-substituting plant	Starts to build an ASEAN-specific model (Asian car) in Thailand	Capability building starts
Not exporting on a large scale	Still not exporting on a large scale	The main mission of the plants is still locally oriented
		Scope of exporting is limited to Oceania, owing to the international division of production

It was necessary for Mitsubishi Motors to expand its production capacity in order to meet the increasing demand for pick-ups in Thailand, Oceania, Southern Europe and so on. However, at that time Mitsubishi Motors' sales in Japan were doing well, though there was little demand in Japan for pick-ups. Mitsubishi Motors decided that it would be better for its operations in Japan to focus on manufacturing passenger cars and recreational vehicles, rather than increasing pick-up production.

For these reasons, Mitsubishi Motors decided to construct an export-oriented plant for pick-ups in the first half of the 1990s and locate the new plant in Thailand, as it was the single largest market for pick-ups after the United States, and it was also a reasonable site for labour and plant location. In 1996, MSC began operation of its export-oriented plant for pick-ups. So, the strategic attitude of the small company, which was aimed at the effective use of its limited managerial resources, emergently led to the strategic decision to construct the export-oriented plant for producing pick-ups (for details, see Orihashi 2000a).

In contrast, TMT constructed a new plant where it largely concentrated the production of its passenger cars, the Corolla and Soluna (an Asia-specific vehicle). Toyota's old plant concentrated its production on Hilux models (pick-ups). TMT expected the demand for passenger cars to increase rapidly together with motorization in Thailand, so the new plant was designed mainly for production for the domestic market. TMT did not engage in large-scale completely built up (CBU) exports, and its position in Toyota's worldwide strategy remained as an import substitution plant.

Production shrinkage stage

MSC started with large-scale exports in 1996, and now it exports to about ninety countries. The main markets are EU countries, especially Spain and Portugal, and Australia. At present, most of the Thai-built pick-ups are exported because of the shrinkage of the domestic market. MSC has had to shut down its old plant, which had been manufacturing pick-ups for the domestic market. Since MSC started its large-scale exports earlier, the company has been able to stay ahead of other local manufacturers that did not start large-scale exports until after the crisis.

The crisis occurred in 1997, when TMT had already finished construction of its new plant and its operations were well under way. Like that of other local manufacturers, TMT's capacity utilization rate dropped as local demand shrank.

Faced with this difficult situation, TMT decided to increase its exports, reduce its human resources and greatly increase its invested capital. Late in 1998, in accordance with a Hilux model change, production of the Australian model moved from Japan to TMT. But TMT had a hard time purchasing local parts that met Australian design rules, so the local content ratio of these export models still remains low. In order to increase its exports, TMT has been working hard to raise the local content ratio of export models and to strengthen its manufacturing capability.

As shown above, in Thailand we could observe a pattern that matched our predictions in this chapter. Japanese automotive manufacturers in Thailand faced a rapid increase in local production continuing up until the Asian economic crisis. Among these, Mitsubishi Motors, the small competent firm, expanded its production on the assumption that it would export worldwide the pick-up model it would produce. This capability to export worldwide gave the firm a buffer so that the crisis had only a small effect. Mitsubishi Motors, having already built up a manufacturing capability for exports, was able to expand its exports smoothly from Thailand with the fall of the Thai baht. We must also take into consideration the fact that MSC already had previous experience of exporting on a large scale, for example in the relatively large-scale export of the Lancer (passenger car) to Chrysler Canada in the first half of the 1990s. MSC managed to overcome the crisis by shutting down its KD plant for the domestic market and integrating this plant's production into its export-oriented plant.

Mitsubishi Motors' strategy, which was influenced by its lack of resources, ended up adapting to the environment after the crisis and has been a source of competitive advantage.

On the other hand, Toyota, the large competent firm with a high dynamic capability, developed a local-oriented vehicle (passenger car) and expanded its local production (including the construction of a new factory for passenger cars) prior to the crisis. These decisions resulted in great pressure on the management of TMT, requiring capability reorganization; TMT needed to gain export capability. This reorganization has been taking place because Toyota possesses a dynamic organizational capability that allows it to cope with unexpected environmental changes such as the crisis.

Although it is only one case of two firms in one industry in one country, this case study is, in general, consistent with hypotheses 1, 2 and 3 regarding a small competent firm and a large competent firm with a high dynamic capability.

Case study 2: crisis due to production reduction and Toyota/Mitsubishi Motors in Australia (Table 7.5)

In this section, we discuss an environmental change that took place in the Australian automobile industry beginning in the mid-1980s and the strategic reaction of the Australian subsidiaries of Toyota and Mitsubishi Motors, namely Toyota Motor Corporation Australia (TMCA) and Mitsubishi Motors Australia Ltd (MMAL) (for additional details, see Orihashi 1998). The drastic change in industrial policy was the cause of the environmental change.

First, a brief outline of the history of TMCA and MMAL up to the mid-1980s is necessary. In the early 1960s, Toyota had a KD export contract with Australian Motor Industries (AMI). From the late 1960s, Toyota started to commit deeply to Australia, taking an equity stake in its Australian operations and then gradually increased it. In 1976, Toyota participated in the plan that enforced 85 per cent local content in passenger vehicles in Australia. In order to meet that quota, Toyota worked hard to establish local production capabilities for its main components. In 1977, Toyota constructed a facility for engine manufacture and stamping in Altona, a suburb of Melbourne.

On the other hand, MMAL was formerly named Chrysler Australia (CAL).

Table 7.5 Toyota and Mitsubishi Motors in Australia

	Mitsubishi Australia	*Toyota Australia*
Capital (in millions of US$)	73.98	481
Shareholder(s)	Mitsubishi Motors (Japan) 60% Mitsubishi Corp. (Japan) 40%	Toyota Japan 100%
Sales (in millions of US$)	2,075 (1997 estimate)	3,900 (1996)
Export vehicles	Approx. 12,000 vehicles (1996)	Approx. 14,000 vehicles (1996)
Exports to	CBU to USA, Japan, New Zealand, etc. Casting product to Japan	CBU and KD to Persian Gulf countries, New Zealand, etc. Some auto parts to ASEAN, South Africa, etc.
Characteristics of the plant	Important overseas location for sales and production, much the same as the USA, Thailand, and the Netherlands	Medium-sized overseas location for production in Toyota's global network
Models	Changed from two models to one model (Magna)	Camry and Corolla
Local content ratio	Approx. 75%	65–70%
Investments after 1984	Investments for model change and rationalization	Construction of new plant, factory reorganization
Employees	5,400 (18 Japanese nationals)	4,220 (20 Japanese nationals)
President	Australian	Japanese

Source: Firm interview (October 1997)

Note
CBU, completely built up; KD, knocked down

In 1971, following a capital tie-up between Mitsubishi Motors and Chrysler Motors Corporation, the companies made a contract covering distribution, trademark and technical assistance in Australia. Mitsubishi Motors started building Mitsubishi vehicles in Australia and the share of Mitsubishi vehicles gradually increased as a proportion of the total production of CAL. Subsequently, Chrysler experienced financial difficulties and asked Mitsubishi Motors to make a financial commitment to CAL. Mitsubishi Motors, together with Mitsubishi Corporation, purchased almost all of the equity of CAL, and CAL became MMAL. Manufacture of the Magna, a model that was an Australia-exclusive modified version of Diamante, was begun and it became a sales hit. Moreover, MMAL developed its own derivative product, the Magna-based station wagon (estate car).

The environmental changes that took place in the Australian automotive industry after 1984 are considered in two stages, based on changes in Australia's industrial policies (Table 7.6). The first was the production expan-

Table 7.6 Analytical framework (environmental side) – Australia

Initially	Production expansion stage (1984–1995)	Production shrinkage stage (1996–)
Conditions		
Local market is limited	Government's new policy (Button Car Plan) forces local suppliers to increase production volume	Local market gets weaker, owing to an increase in car imports, because the government's policy has changed to liberalization
Importing is very difficult, owing to government policy	Minimum production volume (per model) regulation	
Mitsubishi Motors		
Local sales power is relatively weak	Local sales are still not enough to reach the minimum volume, so worldwide production of Magna (Diamante) is concentrated to the plant (except sedans sold in Japan)	Has already started to export to various countries
Import-substituting plant	The mission of the plant changes, but Mitsubishi's investment is limited = do not build a new plant	Has already achieved the capability to face many types of markets worldwide
Not exporting on a large scale	Starting to export on a large scale to various countries	
	Capability building starts	
Toyota		
Local sales power is strong	Increased local sales (including OEM for GM Holden) force Toyota to construct a new, locally oriented plant	Tie-up with GM Holden ends
Import-substituting plant	Make remarkable progress, especially in productivity improvement	Have to begin exports to maintain the operation
Not exporting on a large scale	Still not exporting on a large scale	Capability building starts
		The main mission of the plants is still locally oriented
		The scope of export is limited, owing to the international division of production

sion stage (1984–1992). Until 1984, there was a protectionist industrial policy for the automotive industry in Australia, and local automobile manufacturers supplied high-cost passenger cars for the domestic market. However, in that year a new industrial policy, the Button Car Plan, was announced. It aimed at a rationalization of the model line-ups and production systems of local manufacturers by encouraging competition. The ultimate goal of this policy was a reduction in car prices and an improvement of the quality and productivity of the local manufacturers to enable them to stand up to global competition. The government aimed to increase production per model by reorganizing local manufacturers, with each manufacturer expanding its production scale and reducing the number of its models. Also, the government opened the local market slightly in order to encourage competition. The policy included a penalty for small-scale production of models, so the local firms struggled to form inter-firm tie-ups among themselves, including joint ventures and original equipment manufacturer (OEM) supply.

The second stage of environmental change began in 1992 and can be called the production shrinkage stage (1992–). After the Button Car Plan, industrial policy towards the automotive industry changed in 1988 and again in 1992, as the emphasis of the policy gradually moved from strengthening the local industry to market liberalization. Since the tie-up between manufacturers reduced the strategic freedom of each company, their disadvantages began to outweigh their advantages. The agreements were gradually cancelled, as they turned out to be only a temporary measure against the Button Car Plan.

Production expansion stage

Initially, MMAL considered a tie-up with another local manufacturer, but could not find a partner that met its conditions. It thus became necessary for the firm to survive on its own. MMAL stopped producing small cars in order to concentrate its production on a medium-sized model sold exclusively in Australia. Because Australia was the biggest market for Diamante-based cars, Mitsubishi Motors concentrated its production at MMAL (except sedans sold in Japan) in order to maintain MMAL's production volume. However, MMAL did not make any large investments in exports, other than at its casting plant, whose exports to Japan increased sharply. MMAL managed to export by building the production capability level at the existing facility instead of building a new facility. Doing so created some disadvantages in quality and productivity, but should exports not go well in the future, MMAL would not suffer in terms of its financial resources. In other words, MMAL's strategic decision kept flexibility high.

Toyota decided on a joint venture with GM Holden Australia (GMHA), which had initially come at GM's request. Behind this joint venture lies the fact that Toyota had just entered into a joint venture in the United States with GM. In Australia, Toyota and GM established a holding company with an equal investment by each side. Both companies continued to operate individually, yet provided vehicles to each other on an OEM basis. However, any large-scale investments required the other partner's approval.

TMCA supplied small and medium-sized cars, the Corolla and Camry, to GMHA, and local demand eventually grew to exceed the capacity of the existing Port Melbourne plant. Thus TMCA borrowed capacity from the Dandenong plant, a former GMHA plant, and transferred production of the Corolla to there. However, because both plants were already rather worn down (especially their painting facilities), TMCA decided to construct a new plant in Altona, where press, engine and transmission facilities had already been completed, as is mentioned on p. 127. TMCA concentrated almost all of its production into this new plant, and quality and productivity improved dramatically. At the same time, TMCA pushed forward with a supplier-strengthening programme (initiated in the early 1990s with Australian government support) and moved for progress in building its production capability. As a result, the firm was able to gain competitive advantage over other local manufacturers (for additional details, see Fujimoto 1998b, 2000).

Production shrinkage stage

MMAL did not enter into any tie-ups with other local manufacturers in the production expansion stage and this decision had little effect. As MMAL had already decided, efforts to expand exports continued, with large-scale exports starting in 1992. Most exported vehicles went to the United States. The Diamante Wagon was also exported to Japan.

Why could MMAL not start large-scale exports earlier? In fact, product quality obstructed this option. MMAL encountered various difficulties overseas, especially in the United States, where the Product Liability Act had been passed. Through exhaustive activities to improve quality, the firm's manufacturing capability also improved. However, MMAL's strategic choice not to make a large financial investment affected its production system. For example, because of the small size of its press machines, the Magna's side panel was pressed only after it had been divided into two parts, which were then welded together. At most manufacturers, the side panel is pressed as one part.

As already mentioned, TMCA had worked hard to improve its manufacturing capability. Table 7.7 shows that, given its production size, in 1996 TMCA,

Table 7.7 Australian, American and Japanese parts costs for the Toyota Camry (Japan = 100)

Commodities	Australia	United States	Japan
Coil spring	116	86	100
Outer mirror	112	94	100
Seat belts	109	73	100
Lamps	105	80	100
Tyres	105	91	100
Glass	101	89	100
Average of all commodities	106	96	100

Source: Industrial Commission (1997: 58)

in terms of its cost, was already by no means inferior to manufacturing facilities in Japan and the United States, where the Camry is also manufactured.

TMCA ended its joint venture with GMHA in 1996, owing to decreased demand for the OEM vehicles and further liberalization of the Australian market. As OEM vehicle supply to GMHA ended, TMCA had to export in order to maintain its production volume. Because TMCA had already improved its manufacturing capability, Toyota announced its plan to start large-scale exports from TMCA to the Persian Gulf countries in 1996. In August 1997, after TMCA had completed a full model change of the Camry, large-scale exports to these countries began. However, numerous problems with the quality of these exported vehicles occurred (Orihashi 2000b).

The first problem was with TMCA employees' understanding of the meaning of quality. TMCA employees had the sense that all they had to do was to achieve Toyota's worldwide quality standard. In Japan, on the other hand, Toyota employees produce vehicles that exceed that standard. As a result, and in this sense, Toyota dealers in the Gulf countries thought the difference between 'made in Australia' and 'made in Japan' was a problem. The second problem concerned a lack of skills on the assembly line. The third problem was that the understanding of quality among consumers in the Gulf countries was stricter than that of TMCA's Australian consumers. In other words, these two problems meant that 'made in Australia' did not meet the quality standard of the export markets. The fourth problem was the inferior image of vehicles made in Australia. The fifth problem was the development process. The Camry was a vehicle targeted for worldwide markets, so its development was a joint project between Japan, the United States and Australia. Nevertheless, because TMCA is relatively small in terms of scale, TMCA's facilities were not considered sufficient either at the time of development or when the vehicle underwent a minor change. In fact, Japan and the United States played the main role in the product's development.

TMCA worked hard to overcome the above problems. At first, TMCA formed a repair team and sent them to the Gulf countries. Prior to this, if an imported vehicle manufactured by a foreign subsidiary of Toyota happened to be defective, the local dealer would repair it and send a claim for the cost of the repairs to the subsidiary. In addition, inspection personnel of local dealers in the Gulf countries travelled to Australia at TMCA's expense to inspect vehicles before they were shipped. This project continued until the quality reached a sufficient level. At the same time, TMCA raised the awareness of its employees' sense of quality such that they would regard the Gulf dealers' request as the standard to be reached. Of course, building mutual trust between TMCA and the Gulf dealers was the key solution. As for the development process, Toyota now promotes technical communication among the three countries to a greater extent than in the past.

The problems outlined above caused a change in Toyota's attitude towards international business. Prior to this experience, it was considered wasteful for each foreign subsidiary to have an export market relations function; Toyota's headquarters in Japan performed this function. However, after these problems were solved, TMCA dispatched an Australian manager to the

Middle East for ongoing communication with local dealers in order to prevent new problems from surfacing. Also, in the Thailand case TMT sent a Thai manager to Australia in order to cope with any possible quality problems with the Thai-made Hilux.

With these efforts, TMCA achieved the capability to maintain the necessary high quality standard, superior to that of other manufacturers in Australia. The fact that TMCA exported more than 30,000 vehicles in 1998 allows us to infer that, by engaging in the activities we mentioned above, TMCA has achieved this export capability in a short period.

Discussion

As is outlined above, patterns can be found of firm behaviour in Australia that match the framework and predictions of this chapter, although not as clearly as in Thailand.

When faced with a local crisis that forced a production increase per model followed by a change to liberalization as regulated by the Australian government, Toyota, a large competent firm with a high dynamic capability, viewing its relationship with GM as more important than the domestic circumstances, formed a joint venture with GMHA. This strategy itself was not a simple one-to-one correspondence. As a result, however, TMCA came to provide small and medium-sized cars to GMHA, generating local demand that exceeded the capacity of its existing plant. TMCA thus borrowed capacity from a former GMHA plant at first, and then constructed a new assembly plant in Altona. TMCA consequently expanded its local-oriented production, though this was not at first intended. Needless to say, behind all of this lie Toyota's financial resources, which had been maintained despite economic sluggishness in Japan.

However, owing to the rapid increase in car imports and cancellation of the joint venture with GMHA, TMCA had to change its strategy and expand its exports in order to maintain its operations. This was a great challenge for TMCA but, thanks to the dynamic organizational capability against unexpected environmental changes possessed by Toyota, TMCA managed to start large-scale exports to the Middle East in a relatively short period of time.

We do not know whether TMCA will be able to survive in the long run, but it has been established that its international competence has advanced rapidly through its efforts towards export expansion.

In contrast, Mitsubishi Motors, a small competent firm, selected a strategy that maintained its local production through the expansion of exports. A tie-up with another local manufacturer was not concluded, so MMAL attempted to survive alone. MMAL decided that it could not maintain two models because it was only the fourth or fifth largest manufacturer in Australia. MMAL therefore concentrated its production on one model (the Magna) and began to export it on a large scale in the pursuit of further scale merits. Because Australia was the largest market for the model, Mitsubishi Motors transferred most of its production from Japan to MMAL. Aided by this decision, MMAL completed its transformation into an export-oriented plant.

As compared with TMCA, MMAL started earlier with its efforts to become an export-oriented plant and managed to maintain its operations without any large investments in its production facility. The firm's measures in response to the increase in car imports were less pronounced than those taken by TMCA.

As has been discussed, the strategic reactions of the Japanese automotive manufacturers in Australia towards the industrial policy change were emergent in nature. This phenomenon became a little confusing with the tie-ups between MNEs; however, it is basically consistent with predictions 1, 2 and 3. At first, the large company expands its local-oriented production and then changes its attitude to an export orientation after the crisis; and the small company is export oriented from the beginning and continues to expand its exports after the crisis.

Conclusion

In this chapter, we propose an analytical framework that takes into consideration the differences in scale and organizational capability of each MNE in order to address why the foreign manufacturing subsidiaries of MNEs from the same home country sometimes pursue different strategies in the same host country. As an example that reveals some remarkable differences with regard to this, we examine the strategies of two Japanese automotive MNEs in Thailand and Australia. We have especially sought to measure the firms' resilience during a crisis caused by an external shock, such as a regional economic crisis or rapid increase in car imports. Throughout this chapter, we examine whether our conceptual framework is consistent with these cases.

Needless to say, definitive conclusions cannot be drawn from case studies of only one industry in two countries. However, it is possible to obtain practical results that are consistent with the conceptual framework and the hypotheses we propose. Differences in scale (financial resources) in the home country, manufacturing organizational capability and dynamic organizational capability affect the selection of the type of local manufacturing subsidiary, local product and marketing of the local product, as well as of a firm's response to a crisis. Furthermore, these relationships were found to hold in both the production expansion stage and the production shrinkage stage. This research sheds additional light on the strategic behaviour of MNEs in the 1990s when the global economy began to change rapidly. While respecting the existing conceptual framework of the theory of MNEs and international business, we insist that the individuality of each MNE, for example differences in scale and organizational capability, must be considered more seriously when analysing MNEs.

As this chapter is an exploratory study based on only a few cases, there still remain numerous questions to address. We must study the possibility of generalizing this conceptual framework and these predictions by expanding the scope of the research. Can strategic differences between MNEs based on scale and organizational capability be observed in Toyota and Mitsubishi Motors in countries other than Thailand and Australia? Can the framework be applied to other automotive MNEs? Can consistent findings be obtained in industries other than the automobile industry? Will we need to develop our study to include hypothesis testing by statistical analysis? Of course, we must also continue to examine critically the conceptual framework itself – whether scale, static capability and dynamic capability, as this study showed, are enough to explain the differences in the strategic behaviour of firms.

While many questions, such as those outlined above, remain, we believe the concepts introduced here of 'uniqueness of each firm', 'dynamics' and 'emergence' enrich the analysis of manufacturing MNEs and may be appropriate to some extent. Our conceptual framework has also made a contribution to the literature in explaining some important cases of the strategic behaviour in the automobile industry of Japanese MNEs that were most active in overseas direct investment in the second half of the twentieth century.

Notes

1 The field research by Takahiro Fujimoto took place at Toyota Australia in 1995. The field research by Shinya Orihashi took place at Mitsubishi Australia and Toyota Australia in October 1997 and at Toyota Thailand and MMC Sittipol in June 1999, and again at MMC Sittipol in November 2000. Both authors have also visited the Japanese headquarters of both Toyota and Mitsubishi Motors and are grateful for the cooperation of the companies in this research. The authors would also like to acknowledge the cooperation and suggestions provided by Yveline Lecler of Lyon University and Daniel Heller, a graduate student at the University of Tokyo.
2 Of course, there are some researchers who do pay attention to firm-specific factors (see Kogut and Zander 1992).
3 Rival companies invest in the same area at the same time because of opposition action between oligopoly companies (Knickerbocker 1973).

References and further reading

Abo, T. (ed.) (1994) *Hybrid Factory: Japanese Production Systems in the United States*, New York: Oxford University Press.

Andrews, K.R. (1971) *The Concept of Corporate Strategy*, revised edn, Homewood, IL: Dow Jones-Irwin.

Bartlett, C.A. and Ghoshal, S. (1989) *Managing across Borders: The Transnational Solution*, Cambridge, MA: Harvard University Press.

Clark, K.B. and Fujimoto, T. (1991) *Product Development Performance: Strategy, Organization, and Management in the World Auto Industry*, Cambridge, MA: Harvard University Press.

Dosi, G., Nelson, R.R. and Winter, S.G. (eds) (2000) *The Nature and Dynamics of Organizational Capabilities*, Oxford: Oxford University Press.

Dunning, J.H. (1980) 'Toward an Eclectic Theory of International Production: Some Empirical Tests', *International Journal of International Business*, 11 (2): 9–31.

Fujimoto, T. (1998a) 'Reinterpreting the Resource-Capability View of the Firm: A Case of the Development-Production Systems of the Japanese Auto-Makers', in Chandler, A.D., Hagstrom, P. and Solvell, O. (eds) *The Dynamic Firm: The Role of Technology, Strategy, Organization, and Regions*, Oxford: Oxford University Press.

—— (1998b) 'Toyota Motor Manufacturing Australia in 1995: An Emergent Global Strategy', Discussion Paper, Faculty of Economics, University of Tokyo (unpublished).

—— (1999) *The Evolution of a Manufacturing System at Toyota*, New York: Oxford University Press.

—— (2000) 'Evolution of Manufacturing Systems and *Ex-Post* Dynamic Capabilities', in Dosi, G., Nelson, R.R. and Winter, S.G. (eds) *The Nature and Dynamics of Organizational Capabilities*, Oxford: Oxford University Press.

Ghoshal, S. and Nohria, N. (1989) 'Internal Differentiation within Multinational Corporations', *Strategic Management Journal*, 10: 323–337.

—— (1993) 'Horses for Courses: Organizational Forms for Multinational Corporations', *Sloan Management Review* (Winter): 23–35.

Hofer, C.W. and Schendel, D. (1978) *Strategy Formation: Analytical Concepts*, Cambridge, MA: West Publishing.

Hymer, S. (1976) *The International Operation of National Firms: A Study of Direct Investment*, Cambridge, MA: MIT Press.

Industrial Commission (1997) *The Automotive Industry*, Canberra.

Knickerbocker, F.T. (1973) *Oligopolistic Reaction and Multinational Enterprise*, Cambridge, MA: Harvard Business School Press.

Kogut, B. and Zander, U. (1992) 'Knowledge of the Firm, Combinative Capabilities, and the Replication of Technology', *Organizational Science*, 3 (3): 383–397.

Mintzberg, H. and Waters, J.A. (1985) 'Of Strategies, Deliberate and Emergent', *Strategic Management Journal*, 6: 257–272.

Nelson, R.R. and Winter, S.G. (1982) *An Evolutionary Theory of Economic Change*, Cambridge, MA: Belknap Press of Harvard University Press.

Nohria, N. and Ghoshal, S. (1994) 'Differentiated Fit and Share Values: Alternatives for Managing Headquarters–Subsidiary Relations', *Strategic Management Journal*, 15: 491–502.

Orihashi, S. (1998) 'A Research Study about Strategy Formulation Process in Multinational Enterprises: The Case of the Australian Automotive Industry', Master's degree thesis (unpublished), Graduate School of Economics, University of Tokyo (in Japanese).

—— (2000a) 'Plural Patterns of International Strategy in the Same Industry: The Case of Toyota and Mitsubishi Motors in Thailand and Australia', *Annual Bulletin of the Japan Academy of International Business Studies*, 6: 238–249 (in Japanese).

—— (2000b) 'Breaking from Import-Substituting Plant to Export-Oriented Plant: The Case of Japanese Automotive Makers in Australia and Thailand', *Journal of Asian Management Studies*, 6: 97–102 (in Japanese).

Penrose, E.T. (1959) *The Theory of the Growth of the Firm*, Oxford: Basil Blackwell.

Porter, M.E. (1985) *Competitive Advantage: Creating and Sustaining Superior Performance*, New York: Free Press.

Rugman, A.M. and Verbeke, A. (2001) 'Subsidiary-Specific Advantages in Multinational Enterprises', *Strategic Management Journal*, 22: 237–250.

Sugiyama, Y. and Fujimoto, T. (2000) 'Product Development Strategy in Indonesia: A Dynamic View on Global Strategy', in Humphrey, J., Lecler, Y. and Salerno, M.S. (eds) *Global Strategies and Local Realities: The Auto Industry in Emerging Markets*, Paris: GERPISA.

Taggart, J.H. (1998) 'Strategy Shifts in MNC Subsidiaries', *Strategic Management Journal*, 19: 663–681.

Teece, D.J., Pisano, G. and Shuen, A. (1997) 'Dynamic Capabilities and Strategic Management', *Strategic Management Journal*, 18: 509–533.

Vernon, R. (1971) *Sovereignty at Bay; The Multinational Spread of U.S. Enterprise*, New York: Basic Books.

Womack, J.P., Jones, D.T. and Roos, D. (1990) *The Machine That Changed the World*, New York: Rawson Associates.

Part II

The transferability of 'Japanese' skill formation systems

8 Intellectual skill and its transferability

Kazuo Koike

This chapter highlights the importance of workers' skills as a component of the Japanese human resource development (HRD) systems for the competitive strength of the Japanese economy and discusses the transferability of these skills and systems to other countries. For example, if it were the case that Japanese HRD systems depend heavily on an intense network of low-skilled technical personnel, the systems would not transfer well. However, if the opposite were the case, namely that personnel consist of workers with high levels of technical skills, a high level of transferability could be expected. Thus it is of vital importance to discern the character of workers' skills on the shop floor in contemporary Japanese industry in order to assess the transferability of the Japanese HRD systems.

A variance in workers' skill levels can have a significant effect on efficiency levels, despite the use of similar equipment and machinery. According to a recent series of intensive field studies conducted in Japanese industrial settings, intellectual skill was found to be the variable that most promotes competitiveness (Aichi-ken 1987; Koike and Inoki 1990; Muramatsu 1996; Koike *et al.* 2001). This chapter aims to identify the essential character of intellectual skills, how such skills can be acquired and their prospects under the development of information technology and robotization. Most of the examples presented here have been obtained from results of research conducted in 1998 on twenty-eight workshops[1] in Toyota and its related firms (Koike *et al.* 2001).

This chapter makes use of a qualitative analysis to identify the importance of workers' skills. As far as I know, in the field of economics, no attempt to measure the levels of workers' skills quantitatively has been successful. Through the collecting and analysing of data on significant examples in which skills are being well utilized on the shop floor, an image emerges of the most vital workers' skills constituting the know-how needed to deal with uncertainty on the shop floor.

Whether robotization or developments in information technology depreciate workers' skills is a hot issue. Since the Toyota study also examines workshops that largely utilize robots and information technology, this chapter addresses this question with reference to a couple of cases in the study. Although confined to blue-collar workers, the theory explored in this chapter could be even more intensively applied to professional workers.

The discussion is organized as follows. The first two sections identify the two vital components of intellectual skills needed in dealing with problems and changes, respectively. The third section investigates the impact of technological developments, particularly the heavily 'robotized' workshop, and the implications for these developments on the future. The fourth section is an analysis of the process by which intellectual skills are formed, which is key in assessing transferability of the Japanese HRD systems, and the final section discusses the transferability of the Japanese HRD system.

Dealing with problems

Identifying defective products

A worker's intellectual skill can be defined as his or her know-how, or expertise, in dealing with uncertainty on the shop floor. Uncertainty on the shop floor consists of problems and changes in operations. 'Problems' include issues that are not fully known in advance in terms of their nature, the moment at which they might occur and their size. A most obvious example of a problem is a defect in product quality. We cannot predict exactly what kinds of defects could occur, at what moment and how crucial they are to productivity. If we were able to predict these variables accurately, then a computer program could be designed to identify and handle product defects efficiently.

Changes in production are also uncertain variables in terms of their extent and timing, but their nature is already known in advance. An obvious example is an adjustment in output: demand for products often changes, almost unexpectedly, in both timing and degree. If production workshops cannot effectively adjust to a change in demand, many unsold cars would accumulate in the stockyard, which would be not only an additional cost, but also a waste of scarce resources.

Let me explain in more detail the skill required to identify defective products. The know-how in dealing with such problems is composed first of the ability to find and identify the defective products in the operational flow. On an assembly line in the car industry, for example, the most visible defects are incorrect parts having been attached or required parts not having been attached. Although these are the simplest kinds of defects, they are not actually easy to spot during the usual flow of operations, since the cycle time of a job, or process time, can be as short as sixty seconds in ordinary situations (this becomes longer when markets are slack, as will be seen later). Within such a short period, it is not easy for workers to search out defects that occurred earlier, since they must continue to carry out their own operations.

It could be argued that inspection staff can fulfil this role better than the operators on the assembly line; however, an inspection may cost far more than would be the case if assembly line workers recognized the defect earlier in the assembly process. Consider two very crucial points: first, a defect becomes excessively more difficult to identify further down the assembly line because many parts that have since been attached conceal the original defect; second, even when the defect is identified during later production stages, correcting it requires relatively far more time at that stage, and hence becomes more costly. The simplest defect, such as the attachment of an incorrect part, could necessitate several operations to correct, because many parts that had been assembled after the original defect was made would have to be removed. The consequent damage to productivity would be enormous. Thus it is imperative to locate defects during the performance of the jobs adjacent to the one that caused the defect or, at the latest, within the workshop where the defect occurred.

According to workshop supervisors and veteran workers, the best way to acquire the know-how to identify incorrect or missing parts is to have had experience in working on the preceding assembly-line jobs in the workshop. The reason is clear: to identify the defect in a span of time as short as sixty seconds, knowledge of the normal situation without any defect is indispensable. If an assembly-line worker possesses this knowledge, a simple glance is enough for that worker to identify a flaw. Experience in jobs further down the line also significantly improves an operator's capability to inspect for defects, in that the operator becomes more aware of points requiring careful attention in order to avoid defects. Naturally, then, when a worker with this experience is redeployed in jobs earlier up the line, a minimum number of defects occur. Thus in this case, to acquire the know-how to deal with defects, workers should have a broad experience of most jobs within one workshop.

Identifying causes of defects

Identifying the causes of defects is more crucial to productivity than discovering them is. In the absence of this ability, defects could occur repeatedly. If the causes were not identified and subsequently rectified, machinery would continue to produce defective products, and stopping machinery for fear of producing many defective parts, thereby putting a halt to production, is not a feasible alternative. However, when the operator is capable of identifying the causes of defects and rectifying them, machinery can be corrected so that it does not continue to produce defective parts. The difference in productivity is remarkable.

This kind of know-how requires a higher level of knowledge than simply the ability to recognize that there is a defect. In order to work out the causes of defects, it is necessary to know the machinery's structure and the production mechanisms, because defects can result from any kind of trouble in the machinery or in the flow of production. As the machinery's structure and the flow of production become more complicated with the use of information technology and robots, demands for this knowledge increase.

To illustrate the consequences for levels of know-how when robotization develops, we can consider an example from another assembly-line workshop – one of the largest part suppliers in the Toyota group – that is fully equipped with many robots. Small electric motors are assembled by sixteen workers divided over two shifts together with almost two dozen robots, all under the direction of one supervisor. Automatic machinery and robots carry out most of the assembly work, and the only operations remaining for the workers are those concerned with addressing problems. Dealing with product changes is not such a demanding task here, since robots automatically adjust to changes in the products; a bar code on each product indicates that certain ones are to pass through a specific robot, while others are to be assembled by the other, ordinary robots. When a machine or robot detects something wrong, the machinery stops and a sign lights up to call an operator.

In this case, the ability to deal with equipment such as robots is vital when handling defective products; most defects are due to some kind of equipment trouble. Take, for example, a problem that arises in which products do not flow smoothly at a certain spot. Immediately after the operator in charge of that job has become aware of the slow-down, the operator tries to identify the cause and, if feasible, to rectify the situation before maintenance people arrive. As it takes about fifteen minutes for maintenance to get there, high efficiency greatly depends on the ability of workers to handle problems.

However, a very high level of skill is, of course, required for handling such problems. In order to identify the cause, extensive knowledge of both the mechanical and the electrical dimensions of the equipment is necessary, since a lot of variables are concerned. With regard to the above example, the slow product flow may have been due to trouble on the mechanical side, such as a loose screw, which then prevented products from passing the censor, or the trouble could have originated in the censor itself, or even somewhere else.

To implement full standardization in handling problems requires a high level of skill, since problems differ from one another in character and cause, even in robots made by the same manufacturer. Thus specific knowledge of the trouble history of an individual machine or robot is extremely helpful when trying to detect the sources of problems. In this workshop, if the worker on that particular job finds a problem that is too difficult to solve within five or ten minutes, the worker calls for the workshop's relief operative, the person who is regarded as the most capable in handling problems. After this point, if the problem is not solved within fifteen minutes, maintenance people take over.

Workers are the first to handle robots. When trouble occurs and minute adjustments are required, the workers in the production workshop, rather than the maintenance people, take care of it. This division of labour was a new development, as maintenance people had played the main role in dealing with problems in this workshop only ten years before. The major reason for the change, according to the foreman and the relief personnel in the workshop, was that adjustments required knowledge of specific kinds of products and assembly-line operations, such as whether screw tightening was imperative or not at a certain spot. Without this specific knowledge, minute adjustments were not feasible. Though the maintenance people possessed a greater knowledge of robots, they were not familiar with the specific forms of products and assembly operations.

Dealing with changes

Changes in output

There are four kinds of changes requiring intellectual skills on the part of workers: in output quantity, in production methods, in products and in labour mix. Product demand sometimes fluctuates significantly. If production does not sufficiently adjust to the shifts in demand, the firm's profits will surely be damaged. Yet efficient adjustment on the shop floor is difficult, unless many workers are capable of performing most of the operations in the workshop and some are so skilful that a redistribution of the operations into each job in the workshop is feasible.

The redistribution process is a famous aspect of Toyota Systems, and one that has been widely diffused into other industries. When demand decreases by 20 per cent, for example, Toyota decreases the speed of production by 20 per cent from, say, sixty to seventy-two seconds for making one unit of a car (if the extent of change is smaller than 10 per cent, an adjustment in working hours would be enough to accommodate the change). A simple slow-down in the speed of the manufacturing line results in a cost increase. To prevent this, Toyota tries to reduce the number of workers in a workshop by 20 per cent, say from fifteen persons to twelve, without a decrease in the kinds of operations necessary; if there were a decrease, cars without doors, for example, could appear. Suppose there were sixty operations in the workshop carried out by fifteen workers before, and now those have to be done by twelve workers. The change cannot be implemented unless there are many workers who can carry out many different operations in the workshop. Redistribution would not be feasible simply by allocating 20 per cent more operations to each individual worker, because each operation differs in length of time required and in difficulty, and changes in the order of operations to be handled are unavoidable. Thus intellectual skill is, in this case, defined as a capability for doing many jobs in the workshop.

A more demanding type of know-how is required for the redistribution of operations. It requires knowledge in at least two areas: a thorough knowledge of the features of all the operations in the workshop (level of difficulty and order for assembling); and an awareness of the skill levels of individual members in the workshop who can conduct these operations at any given time. The best people with this knowledge are undoubtedly the veteran workers, as having worked together on a daily basis has afforded opportunities to become acquainted with individual skill levels. This is a very typical example of Hayek's 'specific knowledge'. If, instead, an engineer were to conduct the redistribution, the result would definitely be worse, since the engineer would lack the knowledge that is rooted in the experience of having worked with the operators on a daily basis.

Changes in production methods

Although a change in production methods occurs less frequently than a change in output quantities, its handling requires the most demanding skills. Whereas a change in output occurs several times a year, a change in production methods appears once about every two years when new car models are introduced.

Two skills are required to facilitate this change: the ability to design the procedures for manufacturing or for individual job content, and the ability to have a voice in deployment or even in selection of machinery or equipment. A new car model necessitates a redeployment of the production process, in terms of both the mechanical side and workers' operations. To implement a new production process, Toyota organizes a small preparation team consisting of veteran workers, one per every two or three workshops, along with department engineers for a preparation phase that can last as long as one year. During this period, the team searches for the best equipment, the best deployment of machinery and the best design of operational procedures.

Apart from selecting equipment, the blue-collar team members seem to have a significant voice concerning various items, because, through their daily work experiences, they know best how to deploy machinery in the most effective way and how to design operations. Giving blue-collar workers a large say in deciding these matters, something that textbooks generally advised against, is the most significant contribution of Toyota Systems. It is clear that without this contribution, a less efficient manufacturing process would have long prevailed.

For workers to acquire this high level of knowledge, it is essential that they should experience adjacent workshops that are closely related in terms of the character of skills required. The supervisor has an explicit policy of moving capable candidates to adjacent workshops. These promising workers are required to obtain the skills to identify causes of problems that occur within a couple of workshops.

Changes in products and labour mix

The two other kinds of change demand a lower level of skills than those needed for output and production methods. One is a change in the kind of product. The variety of and shifts in consumer demand require one assembly line to accommodate a diverse range of products. While small changes in the kind of product need few jig or tool changes, others that do require workers to change jigs and tools need far higher skills than normal operations. Skilled workers can change jigs and tools quickly without causing defective products to follow, since minute adjustments are usually crucial during the change.

The second kind of change demanding a certain level of intellectual skills to resolve concerns labour mix. Two cases are illustrative. One is an ability to substitute for absent workers in the workshop. On a continuous assembly line, even one vacant position stops the whole line. Consequently, it is imperative to have workers who can substitute for many positions in the workshop. This necessity is, of course, common to any assembly line in any country. In the United States, these substitutes, called relief or utility personnel, are paid at a slightly higher rate than the others in the workshop.

The other is the ability to teach less experienced workers in the workshop. Newcomers need instruction for even the easiest jobs and thus it is necessary that veteran workers who can instruct them are available. This capability to teach newcomers is an element of workers' intellectual skills.

Four levels of workers' skills

To sum up, two crucial skill criteria have emerged in the above analysis: breadth of work experience and ability to handle problems. Under these two criteria, four skill levels can be identified for assembly-line workers of non-supervisory ranks in the car industry.

Level 1 workers can carry out only the usual operations of one job, simply and without delay, paying no attention to defects. Two groups of workers belong to this level: beginners who have just begun working for the car manufacturer and temporary workers who are seasonal agricultural workers or who prefer short-term employment.

Level 2 includes regular workers with two or three years of experience who can carry out two or three different jobs in the workshop, and are able to identify defects and place red tape at the appropriate points to indicate defective spots. The rapid flow of production and their lack of sufficient skills do not allow them to rectify defects. Even so, indicating the defect is a great help in facilitating repairs at a later stage.

Level 3 workers, whose broad experience covers most jobs in the workshop, can not only identify defects but also understand their causes. They rectify the causes of defects if they have time, so that there is likely to be no recurrence of those particular defects in the flow of production.

Level 4 workers, whose broad experience extends even to the adjacent workshops, can design new operational procedures as well as make decisions about deployment of equipment. They can also function as instructors in overseas plants. Level 3 and Level 4 workers differ little in the skills they utilize within one workshop; both utilize intellectual skills.

Current distribution

I have not yet made it clear how many workers with intellectual skills are needed in the workshop in order to elevate competitiveness. If only a few workers of Level 3 and 4 are required among many Level 1 workers, the fostering of intellectual skills may not be particularly crucial for technology transfer. A step towards a reasonable estimate of the necessary number of workers with intellectual skills is to identify the current distribution of the four skill levels. According to the results of a Toyota study that covers various occupations of blue-collar workers, it is natural that die-making or maintenance workshops consist mostly of Level 3 and 4 workers, with a few others still in their training period. Yet the focus here is on those assembly-line workshops that seem to have the least need for high-level skills. The Toyota study shows that in an ordinary assembly-line workshop, Level 3 and 4 workers make up for nearly 50–60 per cent of the total, while Level 1 and 2 workers constitute 20–25 per cent and 20–30 per cent respectively.

Since the survey was conducted during a year of slack labour markets, it is possible that an excess of workers with high skill levels was found. Furthermore, no temporary workers actually worked in the workshop that year. Instead, owing to an unexpected decrease in demand for various car models, temporary transfers of regular workers from other workshops made up nearly 20 per cent of the composition of workers. A glance at labour distribution when labour markets are tight may help explain the over-investment in Level 3 and 4 workers. According to the statements of supervisors, the maximum distribution of the Level 1 workers for the workshop at the time the survey was conducted was a quarter, and that does not differ much from the current figure.

This figure can be supported by surmising the incidence of distribution variation within three known parameters. First, if the number of Level 1 workers greatly increases, many defects are overlooked, and the cost of identifying and rectifying defects at the end of the assembly line becomes too high. To prevent this, it is crucial to assign each veteran worker to a position immediately after the job being performed by a Level 1 worker. Second, not all jobs in the assembly-line workshop are appropriate for Level 1 workers. Even seemingly easy jobs require a high level of skills, such as the skill to identify any uneven surface or any gap between doors and pillars. Third, the skills of veteran workers of Level 3 or 4 cannot be developed instantly; it takes time to build up their skills. Reaching Level 2 is an indispensable step on the way to reaching Levels 3 and 4. Taking these parameters into account, the current percentage of Levels 3 and 4 workers, 50 or 60 per cent, is not likely to be very different from the optimum.

The percentage of Level 4 workers will be determined by at least three factors: in terms of employment, the ratio of overseas plants to domestic ones; the frequency of model changes; and each plant's stage of development. All these are changeable, and the last factor, for example, requires a higher ratio during the initial period or the phase during which a new car model is launched, and a lower one as a plant becomes older. If we remember that their successors, including the preparation staff, need to be trained, the proportion of Level 4 employees is likely to differ little from the current figure. It is worth mentioning that the above picture applies only to blue-collar workers. Engineers constitute another large part of overseas staff.

According to other in-depth studies of various industries, this high distribution of intellectual skills is not confined to Toyota, but is widely diffused throughout many workshops in large firms in contemporary Japanese industry. Yet it should not be assumed that these high levels characterize most workshops in Japan. The smaller the firm, the lower the percentage of workers with intellectual skills. A rough estimate would be that workshops with many highly skilled workers are in a minority.

The impact of technological developments

Robotization

Will this ratio increase or decrease as a result of intensive robotization or the development of information technology? To answer this question, an investigation of workshops that are highly robotized or extensively equipped with information technology is necessary. The Toyota survey covers an assembly-line workshop that is fully robotized, and a die-making workshop that utilizes computer-aided design and manufacturing (CAD/CAM) to the fullest extent.

Apart from the support jobs, the workshop is fully equipped with robots, which clearly indicates that the mainline jobs are occupied only by intellectually skilled workers. The reason for this is evident: as the repetitive operations are wholly conducted by robots, problem handling remains in the hands of workers. Since robots, as well as machinery, are more complicated in their structure than would be the case in a less well-equipped plant, it would hardly be feasible for workers without intellectual skills to conduct the mainline work. For the peripheral jobs, such as distributing materials or packaging, workers with lower-level skills are deployed; temporary workers or inexperienced regular workers are expected to learn the mainline work as ad hoc helpers on the main line.

Die-making workshop

Another indicative case is a workshop for fitting and erecting complicated dies. Needless to say, in die making, utilizing hand tools for grinding is one of the most demanding jobs in terms of skills required in the traditional sense. Yet the extremely high skills required have largely transformed the job from an exercise in manual dexterity into one that is more intellectual in nature.

A shift towards the use of more intellectual skills is particularly recognized in two senses. First, in terms of traditional skills, the intellectual component becomes more important, while manual dexterity becomes less so. The introduction of extremely accurate machine tools with electric discharge along with the use of CAM greatly decreases the need for manual dexterity in fitting and constructing, although needs for dexterous fitting or proficiency in machine tool operations do still exist.

A noteworthy example of this is the level of know-how needed to have a voice in designing dies. In Toyota plants, when design engineers determine the framework of new dies, there is an opportunity for the veteran blue-collar workers to comment on the new design. They could judge it inappropriate if they expected many problems to crop up when it came to be manufactured, or if it might cause defects such as burrs or unnecessary fringes, whose correction would need much additional work.

Interestingly, the design engineers listen carefully to the comments of the veteran blue-collar workers, for it is they, rather than the capable design engineers, who know most about how to manufacture dies. Yet this level of cooperation can be maintained only when the opinions of the blue-collar workers turn out to be valid, which is easily tested through the actual manufacturing process. The blue-collar workers must have an in-depth understanding of the die structure, practical knowledge and also experience both of manufacturing dies and of the problems that arise when dies are utilized for production.

The second sense in which the shift towards the use of more intellectual skills can be recognized can be seen in the numerical design process and production workers who engage in fitting. As has been repeatedly emphasized thus far, spotting defects and identifying their causes are essential intellectual skills of workers. This reasoning capability is required to extend to the continually progressive sphere of information technology, as clearly indicated by the case of a production worker in the die-fitting workshop. The design process, which naturally utilizes CAD, is divided into two phases: the conceptual and the numerical stages of design. The former is concerned with the structure of dies, in which the veteran blue-collar workers have a voice. The latter provides numerical data for the design. A worker engaged in die fitting can deduce that a particular defect in fitting is due to some miscalculation in the numerical design process. Unless the worker had taken part in the numerical design process for some time, such reasoning capability would not be possible. Although this observation pertains only to some of the workers currently in the workshop, those who have some knowledge of numerical design could be better at identifying the cause of defects in die making. This type of career extension had rarely been conceived of before information technology arrived on the scene.

The above examples clearly show that the skills required are becoming highly intellectual in nature. The fundamental question, then, is: how are intellectual skills fostered? It is easy to imagine that because the skills required these days are so intellectual in nature, Japanese blue-collar workers tend to be college graduates. In fact, almost the reverse is the case, as is explained below.

Formation of intellectual skills

Grade of education

Most blue-collar workers in contemporary Japanese industry have twelve years of schooling. Exceptions to this are Toyota and its group, in which there are two groups among those with twelve years of schooling. The main group is composed of workers who graduated from ordinary senior high schools at the age of 18, which is no different from those in other large firms. In Toyota, the other, smaller group is composed of workers who finished three years of full-time schooling fully sponsored by the company, after nine years of compulsory education.

These schools, fully sponsored by companies and usually called 'trainee schools', had been common among many large firms in Japan during the period from the First World War to the 1960s. The most famous examples are Hitachi, Nippon Steel and many other established companies. Since the mid-1960s, most boys and girls have progressed to senior high schools in the expectation of finding better job opportunities, and these trainee schools have been in decline owing to the difficulty of attracting boys with high aptitudes. Nowadays, among the few companies maintaining trainee schools are Toyota, Denso and its relevant group, along with Hitachi.

Yet the ratio of trainee school graduates to regular workers has always been small. Most blue-collar workers discussed in this chapter have twelve years of schooling with little technical training before entering Toyota. Thus the core method of forming intellectual skills is principally 'broad and in-depth OJT' (on-the-job training) on the shop floor, supplemented by short periods of Off-JT (off-the-job training).

Broad, in-depth OJT

By OJT, it is implied that a worker experiences most jobs in one workshop for as long as ten to fifteen years. The practical method adopted by these workshops in the Toyota study is not so-called regular rotation, as is usually thought typical of Japanese systems. Instead, irregular rotation is more common. After about one year on one job, workers would be asked to move to another job in the same workshop in order to expand their work experience. Around five years later, a division into two groups emerges from among the workers: those in one group stay in a small number of jobs for a long time while those in the other continue to move frequently. Moreover, some members of the latter group later move to other workshops that are closely related in technology, such as adjacent workshops on the same assembly line.

In order to let a worker move to other jobs in the same workshop, a sub-foreman or a veteran worker, who is partly engaged in his own line job, usually acts as instructor. This instruction takes the following stages: first, the instructor shows how to perform the job; next, the trainee does his own job while being observed by the instructor; and finally, the instructor goes back to his own job, while the trainee continues by himself, referring to the instructor with questions only when problems arise. The period taken for the first two instruction stages is as short as several weeks, while the third stage might last a long time, particularly when the instruction extends to know-how in dealing with problems and changes. Through this broad OJT, the worker is also exposed to the handling of problems in that job. This is in-depth OJT through which intellectual skills are built. Exchanging information obtained through handling problems with fellow workers is another way of promoting intellectual skills.

Short Off-JT

To develop know-how in dealing with problems, it is necessary to acquire basic knowledge of the structure of products and of the theory of electricity or of machining. Without knowing the structure of a car, for example, it would be almost impossible to identify the cause of defects in it. This knowledge is best acquired by short Off-JT, mostly in-house in the case of large firms. These Off-JT courses are well diffused among the members of the workshops in the Toyota study.

Yet it would be hard to say that Off-JT plays a key role in forming these intellectual skills. It is basically because Off-JT courses are appropriate for teaching those items that are standardized in handling that Off-JT courses are also standardized, but, as has been repeatedly emphasized, problems are not fully standardized in their character, timing and extent. Thus it is hard for Off-JT to play more than a supplementary role.

Off-JT courses do not usually last longer than two days and are inserted in between OJT on the shop floor. Thus we may call them short, inserted Off-JT. There are more Off-JT courses for workers in workshops that deal with intensive technology, such as robotization; these courses are clearly more frequent and cover more topics, so that in aggregate more of the workers' time is devoted to Off-JT. A simple conclusion could be drawn that the higher the technological development, the greater the weight of Off-JT courses (for more detail, see Koike 1997).

Pay-for-job-grade

For building intellectual skills, appropriate incentive systems are indispensable. Appropriate incentives need to assess the extent of skill development fairly and to provide an appropriate reward. Yet assessing the development of intellectual skills is not easy and can hardly be done by looking at the single job in which a worker is currently deployed. Suppose there are two workers who are currently assigned to similar jobs in the same workshop. Also suppose that one of the two workers is capable of conducting almost all the jobs in the workshop and accordingly can substitute for absentees, instruct inexperienced workers and, in particular, handle problems, while the other cannot perform these operations, having had no experience in any job in the workshop other than the current one. Under pay-for-job systems, these two workers are paid equally, so that no extra incentive is offered to the one who can deal with problems. For that reason, pay-for-job systems are not appropriate for encouraging the acquisition of intellectual skills. Rather, systems that can evaluate problem handling and the breadth of experience, such as pay-for-job-grade systems, are urgently needed.

Pay-for-job-grade systems should have four vital features: job-grade systems, range rate, yearly increments and merit rating. These features work extremely well in fostering the acquisition of intellectual skills. First, job grades can reflect the level of intellectual skill by considering both the breadth of experience and problem handling, instead of just the single job in which the worker is currently deployed.

Second, the range rate can reward for skill development while a worker stays on one job, which is commonly recognized to occur particularly for the initial three years or so on a job. For example, an industrial relations manager is much more skilful in dealing with labour unions in the third year than in the first year. If this skill development is not assessed, few are encouraged to develop their skills on the job.

Third, performance appraisal is necessary for assessing the individual level of skill development when a high level of skill is required. Naturally, differences emerge between individuals in the levels attained. If these differences were not assessed and rewarded, few would make the hard effort required to enhance their skills.

Fourth, the long term is needed for forming intellectual skills, since broad experience, the essential part of intellectual skills, necessarily takes time. Workers need to be encouraged to stay in the company for a long period. Regular yearly increments in pay are clearly an effective incentive.

Pay-for-job-grade systems with these four features are in operation for both blue-collar and white-collar workers in large firms in contemporary Japan. At a glance, this statement may sound curious when we remember the popular opinion that Japanese pay systems are largely based on seniority. Since this issue is discussed in detail elsewhere, it will be touched upon only briefly here. Although it is true that pay systems for workers in large firms are seemingly composed of various components other than the pay-for-job-grade, an in-depth study has revealed that the main part of these various components is, in practice, decided by the job grade (Koike 1994).

All four features referred to above are also common among ordinary pay systems for professional workers in large corporations in Western Europe and the United States. This implies that blue-collar workers in large, present-day Japanese corporations are paid under renumeration systems practically the same as those for professional workers in the West. In other words, this is 'white-collarization' of the blue-collar workers, as discussed in my book (Koike 1988).

Transferability of the Japanese human resource development system

Larger transferability of broad experience than of problem handling

Transferability of the system needs to be examined for the three crucial features of the system discussed here: broad work experience in one workshop, problem handling on the shop floor and incentive plans to encourage the fostering of intellectual skills. Only a few case studies are available that examine the issue of transferability, since work experience as well as problem handling can be disclosed only by intensive fieldwork that is not easy to conduct. Below are some conclusions drawn based on these few studies.

One particular study (Koike 1998) investigated these three crucial features in a comparison between NUMMI and its prototype plant in Toyota. Although it did not deal with Asia–Japan joint ventures, other studies too have rarely examined the issue on these three vital dimensions. The study found that a broad experience was the most highly transferable aspect of workers' skills, while skills in problem handling were far less transferable, and incentive plans rated lowest of all in terms of their transferability. More concretely, the study revealed that NUMMI workers with short service records have broader work experiences, in terms of the variety of jobs in which they have been engaged, than their Toyota counterparts; however, in terms of their breadth of work experience, Toyota workers surpass those NUMMI counterparts with longer service records.

A look into why this tendency exists leads us to broader conclusions. The policy of stimulating a varied work experience is easiest to pursue, with few institutional hazards, provided the work is confined to one workshop. Because a fraction of this policy had already been implemented in several workshops throughout various countries, such as in the form of relief or utility personnel substituting for absentees, its benefits have thus far been partly realized. Extending this policy beyond a small minority to more workers is one step further towards fostering intellectual skills. Additionally, as already mentioned, achieving a broad work experience can result in having greater mobility between jobs within a workshop. The jobs in one workshop usually fall under the same occupation and, accordingly, demarcation is hardly an issue.

The seniority system presents difficulties for implementing egalitarian rotation to all jobs in a workshop, but a possible solution would be to implement rotations within the same grade, which would render egalitarian job rotation and the spirit of seniority compatible. Both approaches are intended to prevent rotation being arranged according to the discretion of the supervisor.

The same reasoning can be applied to the observation that there is far less problem-handling transferability. Handling problems on the assembly line, as opposed to off-line handling, such as quality control activities or suggestion plans,[2] is a violation of traditional demarcation regulation between assembly-line workers and maintenance people. In other words, each belongs to a different trade. It has become apparent that assembly-line workers have become reluctant to deal with problems on the assembly line because of the complaints of the maintenance people. This demarcation discourages assembly workers from handling problems on the line, despite the fact that management formally encourages them to do so.

In this case, incentive systems serve only to strengthen this demarcation. Even if assembly workers have acquired the necessary skills to solve problems, they can expect no resulting increase in pay under the current pay schemes. Apart from the initial two years, no pay increase is stipulated unless either an overall wage increase is instituted or the worker gains promotion to team leader. That said, however, the pay increase one receives when promoted to team leader can be as little as 3 per cent, an amount that seems too insignificant to encourage workers to acquire demanding skills. Furthermore, unless employment is rapidly growing, such as during the initial start-up period of a plant, there is little chance of being promoted to team leader. In other words, the prevailing policy dictates a single-rate plan for the majority of jobs in the plant, and a slightly higher single rate for maintenance people. Because of this, no incentives are in place that can be expected to promote intellectual skills effectively.

This is not meant to imply that this practice will not change. Pay-for-knowledge, or pay-for-skill, plans now diffusing to some extent throughout the United States. clearly show that pay-for-job can be converted to a pay plan that assesses the breadth of work experience. Yet this pay-for-knowledge plan normally enforces a pay-rate ceiling within a period as short as five years (Ledford 1993), too brief to acquire intellectual skills and too brief by far when compared to the longer period of time that ordinary large Japanese firms expect for skill development in blue-collar workers. The five-year pay-rate ceiling is most likely due to workers' hostility towards merit rating or favouritism. Yet when workers are required to attain a high level of skills in order to deal with difficult problems and changes, inevitable differences emerge between individual workers with regard to the level attained or the time taken to reach the highest level; it is imperative that these differences are assessed by supervisors who are well acquainted with the work. This kind of assessment is merit rating, and because workers dislike this subjective merit rating, the period to the pay-rate ceiling must be short.

Transferability to ASEAN countries

The above arguments could also be applied to transferability of the Japanese HRD system to South-East Asian countries, where a higher level of transferability would be expected than in the West, mainly because of the remarkably different tradition of industrial relations. In the West, there is the least amount of emphasis pertaining to the blue-collar workers in merit rating, thus, to a certain extent, restraining opportunity to acquire higher levels of skill. Merit rating is practised far more in South-East Asia, where the tradition of industrial relations is still young. Thus in South-East Asia, demarcation is a small issue and minimal antagonism against merit rating can be seen in the blue-collar workers. This could be a favourable foundation for fostering intellectual skills, provided the relevant actors in HRD are eagerly willing to do so.

This alone, however, can never guarantee a promising prospect for the transfer of the Japanese HRD system, as South-East Asian human resource management practices do not seem very eager to adopt it. This is mainly because they do not recognize this system as the vital route towards effective skill formation. Rather, they tend to prefer the classic textbook approach; that is to say, workers' skills are considered to be best enhanced when fully based on formal Off-JT. Such an approach neglects the importance of work experience on the shop floor. One recent example of this policy concerns the manner in which foremen are appointed: preference is given to those with higher grades of Off-JT certification or education, but without much work experience. Thus even though there would be ample space to accommodate and even improve intellectual skills, prospects are not promising.

Misunderstanding in contemporary Japan

A similar misunderstanding is rapidly growing even in contemporary Japan. Many actors involved in human resource management in Japan are now quickly changing their systems from that described in this chapter to one that stresses short-term performance. Instead of pay-for-job-grade plans that emphasize long-term performance and hence accommodate skill development prospects, quite a few firms are now eager to convert their system to pay-for-job plans with merit rating, or pay-for-performance in the short term. This development started for managers or non-union members, and is now starting to spill over into labour union members. The prolonged depression, particularly in the banking industry, is no doubt one of the motivating factors for the change. The perception that pay-for-performance for the short term is in common use throughout most large American firms is probably the other motivating factor for this change. In other words, a lack of sufficient intellectual investigation and analysis might be a crucial defect in the Japanese business and academic arenas.

Apart from these perceptions, the Japanese system itself is not free from problems. One of them is the mingled character of pay plans. As stated above, basically, Japanese pay plans are pay-for-job-grade. Yet in practice, many other components that are attached conceal this basic nature. It is because of this mingled character that Japanese pay plans are thought to be seniority based, and hence an eagerness exists to change the pay plans to those measured against short-term performance. It is of prime importance, therefore, that Japanese firms purify their pay systems to pay-for-job-grade by excluding other attached components.

Although contemporary Japanese systems suffer from problems, the ones applied to blue-collar workers concerning pay systems or demands for high levels of skill seem to coincide, at least partly, with the essential character of HRD systems as applied to professional workers in the West. There is no space here for a detailed explanation, but this feature is analysed by Koike (1988) as the 'white-collarization' of Japanese blue-collar workers. If this analysis is correct, prospects for the successful transfer of the Japanese system could be substantial, considering that the number of white-collar workers is increasing in many countries.

Notes

1 By the word 'workshop', I imply a group of workers organized under one foreman, usually consisting of ten to thirty people. The foreman is defined here as the lowest-ranked supervisor, who is mostly working off the line.
2 This would be a common perception, as a recent study on the Japanese car industry suggests. This study insists that Japanese workers deal with problems only while they are off the line, such as in quality control circles or with suggestion plans, and that their on-line activities are fully standardized and do not allow any discretion in dealing with problems (Katz and Darbishire 2000).

References and further reading

Aichi-ken (1987) *Intellectual Skill Formation*, Nagoya: Aichi-ken, Rodo-Bu (in Japanese).

Japan Institute of Labour (1998) *Human Resource Development of Professional and Managerial Workers in Industry: An International Comparison*, Tokyo: Japan Institute of Labour.

Katz, H. and Darbishire, O. (2000) *Converging Divergences: Worldwide Changes in Employment Systems*, Ithaca, NY: Cornell University Press.

Koike, K. (1988) *Understanding Industrial Relations in Modern Japan*, London: Macmillan.

—— (1994) 'Learning and Incentive Systems in Contemporary Japanese Industry', in Aoki, M. and Dore, R.P. (eds) *The Japanese Firm*, Oxford: Oxford University Press.

—— (1997) *Human Resource Development*, Tokyo: Japan Institute of Labour.

—— (1998) 'NUMMI and Its Prototype Plant in Toyota', *Journal of the Japanese and International Economies*, no. 12.

Koike, K. and Inoki, T. (1990) *Skill Formation in Japan and Southeast Asia*, Tokyo: Tokyo University Press.

Koike, K., Chuma, H. and Ota, S. (2001) *Workers' Skills in Manufacturing Industry*, Tokyo: Toyo Keizai.

Ledford, G.E. (1991) 'The Three Case Studies in Skill-Based Pay', *Compensation and Benefits Review*, March/April.

Muramatsu, K. (1996) 'Intellectual Skills in Mass Production Industry: Integrated Systems versus Separated Systems', *Monthly Journal of the Japan Institute of Labour*, no. 434 (in Japanese).

9 Skill formation in the Thai auto parts industry

Nipon Poapongsakorn

The automobile industry in Thailand has undergone a process of rapid expansion and modernization as a result of rapid economic growth since the late 1980s, policy liberalization in the early 1990s and intensifying competition among a few global multinational car-making enterprises (MNEs) (Romijn *et al.* 2000). Since the industry is highly intensive in terms of skill, technology and scale, there is a need for the MNEs to develop technological capabilities and organizational techniques not only in their car-assembly subsidiaries, but also in the companies supplying them with parts. Both capabilities require human resource development (HRD).

The rapid expansion of the automobile production industry resulted in a large increase in the number of jobs, particularly for skilled workers. In response to the smallness of the pool of skilled labour, the Japanese automakers and a few local parts suppliers had to invest in skill training. This chapter will discuss the role of the Japanese MNEs in this HRD, focusing mainly on the process of skill formation and skill upgrading in the Thai automotive industry.

The Asian economic crisis in 1997 was a serious blow to the automotive industry in terms of both car production and employment. However, two interesting observations can be made: first, the effect on output was much more severe than the effect on employment; and second, the negative impact on training seems to have been only temporary. This chapter addresses both issues and argues that it is in the interest of the Japanese investors to protect their investments in Thailand.

Training incidence declined during the deep recession in 1998, but has rapidly picked up again since 1999. Using a survey conducted among 298 workers in four auto partmaker firms, this chapter will attempt to address the issues of determinants of training and its effects on workers' earnings. The analysis will cast additional light, I hope, on our understanding of the skill-formation process in the Thai automobile industry.

This chapter starts with a brief description of the industrial restructuring in the Thai automobile industry. Following this is a discussion of technological development, including a comparison of the effects on employment during the two periods of rapid economic growth and crisis. Then the role of the Japanese MNEs in skill development is described and the process of skill upgrading is analysed. There follows a quantitative analysis of training and workers' earnings. Finally, some conclusions are drawn and policy implications are offered.

Technological development in the automobile industry

Technological development in the Thai automobile industry is heavily influenced by the Japanese carmakers. Changes in their strategies in response to policy environment and changing competition have affected the nature of both technological development and technology transfer in Thailand. Two major types of technological development have been adopted by the carmakers' affiliates, as well as the parts suppliers, namely, production technology and organizational management. After both types of technological development have been described, technology transfer will be discussed.

Technological development

During the 1970s and 1980s, when the assembly plants were very small, labour-intensive techniques were adopted. Thanks to the flexible local content policy, parts produced were simple, low-value products, using labour-intensive techniques (Doner 1991). The Japanese organizational management system was also introduced in the Japanese assembly firms and Japanese parts suppliers. They include *kaizen, kanban,* quality control circles and, later, total quality management (TQM).

Rapid economic growth in ASEAN and the establishment of AFTA, the ASEAN Free Trade Area, in the early 1990s attracted non-Japanese automakers to the region. As a consequence, Japanese carmakers were forced to change their strategy in ASEAN. In a rush to maintain their competitive advantage and market share, the Japanese carmakers expanded and added new production capacity in ASEAN, particularly in Thailand. They also persuaded a large number of Japanese parts suppliers to establish plants in Thailand. During 1986–1991, thirty-seven new Japanese supplier firms were set up. In addition, twenty-eight Japanese suppliers also signed technological assistance contracts with Thai firms. Not only were new plants larger, in order to exploit the scale economies, but production techniques were more capital intensive. For example, the new Toyota plant in the eastern province of Thailand employed robots on the assembly lines. More sophisticated parts requiring higher technology were also produced in ASEAN. To reach a size at which scale economies applied, the Japanese MNEs also introduced the following practices.

First, although Japanese assembly affiliates in ASEAN adopted the *keiretsu* system of long-term relationships, or the cluster network, with the parts suppliers, the system was adapted to allow partmakers to supply their products to more than one carmaker. For example, Toyota and Honda in Thailand procure door stamping parts from the same supplier, while Siam Toyota supplies cylinder blocks to Toyota and other carmakers (Fujita and Hill 1996). Most carmakers are also members of a trade association, which is used as a means to communicate information and to provide training for parts suppliers.

Second, at the regional level, Japanese carmakers established ASEAN regional production networks. As Japanese MNEs turned Asia into an altern-ative site for export-oriented manufacturing, this had to progress according to technology levels in the ASEAN countries. It should be noted that the web network is a development from an existing cluster network built in the 1970s and 1980s (Hatch and Yamamura 1996). The latter was necessary for cost reduction as assemblers began to make local production possible.

Both developments, as well as the increased production capability of the affiliates and local parts firms, have allowed Japanese assemblers in the region to cut costs and to increase cooperation in designing both with their affiliates and with parts manufacturers. For example, in the mid-1970s Toyota allowed the Thai R&D team to work with a Japanese counterpart in designing the Toyota Soluna car. One Thai parts manufacturer co-designs car bumpers and door structures with Isuzu. The Thai Rung Union research department has been engaged in designing the body of a multi-purpose vehicle. Sammitr Motor, a parts manufacturer, has implemented its own R&D for the replacement equipment manufacturer (REM) steel roof market. Stanley Electric, for its part, intends to make its Thai operation its core plant in Asia. It will supply dies and parts to the company's plants in India and Vietnam. For its die-making facilities, the Thai plant has an R&D centre and an integrated production system for automobile lamps; it is thus forced to acquire or develop its own technology. Meanwhile, those that are technologi-cally deficient join forces with sophisticated parts manufacturers from developed countries. Toyota announced its intention to have more parts made in Thailand and other Asian nations.

Third, advances in information technology that are facilitating materials sourcing will certainly have a significant impact on the automotive industry in Thailand and ASEAN. In addition to bringing lower transaction costs for trade in goods and services and reduced costs of parts and components, e-commerce will allow local automotive suppliers to enter overseas markets.

While American automakers and component suppliers are increasingly using Internet links to cope with changing global trends in procurement, the Japanese automotive manufacturers and their partners in ASEAN are still lagging behind their American competitors (Findlay *et al.* 2001). Previous electronic interchange data systems (such as CALS, or commerce at light speed) used in the Japanese automotive industry network jointly developed by Toyota and the Ministry of International Trade and Industry (MITI) to link parts producers in Asia were expensive compared to the Internet. Now, an increasing number of part suppliers both in Asia generally and in Thai-land are considering more procurement through e-commerce systems. For example, steel manufacturers in Thailand will establish an e-marketplace in 2001 (*Auto Asia*, 21 December 2000). Currently, most e-commerce activities are for automobile sales. In 2000, Toyota Motors Thailand set up a Toyota business reform system for its ninety local dealers to speed up communica-tions with dealers (*Prachachart Business*, 3 February 2000). GE Capital (Thai-land) Ltd has formed a strategic alliance with the Central Retail Corp. and car manufacturers Nissan, BMW and Mitsubishi to launch products online.

Thailand's Senior Com also developed a new B2C e-market by providing the hire-purchase software for the car industry (*The Nation*, 30 November 2000).

The Thai government has also provided support for e-commerce ventures, e.g. a Board of Investment's B2B site for Thailand's automotive and electrical parts manufacturers in mid-2001, and an Internet site to provide a wide selection of e-commerce transactions in the automotive industry.

Fourth, some assemblers and parts manufacturers have established regional coordination centres to reduce risk and improve the efficiency of ASEAN operations, a move designed to prevent repercussions from the 1997 financial crisis (Mori 1999). There are plans to use Bangkok as a centre for marketing in the Indo-China region. Chrysler Sales and Services (Thailand) has announced plans to open a new head office, showroom, service centre and parts warehouse in Bangkok in 1999 to be the marketing hub for ASEAN. Japanese assemblers have likewise established regional headquarters in Singapore (Toyota) and Thailand (Honda, Mazda) to facilitate the smooth operations of their networks.

Fifth, quality control and product standardization are becoming more crucial for parts manufacturers. Japanese auto assemblers adopted the American part defect rate control measured in terms of parts per million (p.p.m.) in 1998. The aim is to improve productivity by reducing defect rates. Moreover, since the ISO has become one of the most important quality indicators of exported products, major export markets, such as Europe, North America and ASEAN, require automotive and auto parts manufacturers to be ISO 9000 and ISO 14000 certified. QS 9000 certification now plays a crucial role in the automotive industry, particularly for small and medium-sized enterprise (SME) suppliers that need to enter and sustain participation with American carmakers. Many automotive manufacturers are now taking appropriate steps to achieve certification.

Technology transfer

There are two main channels for technology transfer in the auto parts industry: through business ventures with foreign auto part suppliers and through technological assistance. First, the business venture with foreign auto part suppliers has played an important role in technology transfer since the late 1970s. During that period, when the Thai automotive industry was under a progressive scheme to promote local industry, it became necessary for foreign automobile manufacturers to invite their long-term partners who produced more sophisticated parts to invest in Thailand. At the same time, local suppliers were also in need of technology upgrades of their production processes. When these foreign partmakers formed business ventures with local companies, local suppliers adopted the original production techniques and technology, completing the first stage of technology transfer. Kato (1992) concluded that in this respect a transfer of production engineering technology has been moderately successful, whereas in other areas, such as design technology, progress has been relatively limited.

Beside such links, auto assemblers usually provide essential technical assistance and training to their local suppliers, as it is less costly for these manufacturers to use some locally made parts. For instance, Thai Rung Union Car, the Thai manufacturer of multi-purpose Isuzu pick-up trucks, contends that it received technology transfer and expertise from Isuzu through technical assistance.

Many automakers have established cooperation clubs with part suppliers to provide them with technical assistance, as well as to conduct training. In general, forty to fifty suppliers in Thailand take part in a cooperation club.

A number of Thai partmakers have also received technical assistance from Japanese parts manufacturers. Between 1966 and 1991, forty-two Japanese manufacturers provided technical assistance to Thai firms (Poapongsakorn 1997: table 2.6).

In addition to these main methods of technology transfer, other channels through which Thai parts suppliers gain new technology are, for example, from foreign customers of their exports in REM parts, and also through experiencing the competition of the world market. On the home front, spillovers occur when employees of firms that have acquired technology resign and join other local firms, and when updated knowledge is required for the use of new machines.

Both the Japanese and the Thai governments have also played important roles in technology transfer in the Thai automobile industry. Unlike its old economic cooperation policy in South-East Asia from the 1950s, which was based on the need to secure a steady supply of raw materials and a low cost production base for some labour-intensive industries, the Japanese policy shifted ground after the dramatic appreciation of the yen in the mid-1980s (Hatch and Yamamura 1996: 118). Threatened by the loss of its international competitiveness, Japanese industry began to see the region as an extension of its home base. It therefore pressured the government to promote the new international division of labour in which East and South-East Asian economies were integral parts of the Japanese regional production network. In the view of one Japanese government researcher, the new division of labour would not only benefit Japan, but also promote the expansion and maintenance of Asian economic growth (Kitamura Kayoko, quoted in Hatch and Yamamura 1996: 119).

The Thai automobile industry received significant assistance from Japan through various programmes. In 1988, the Japanese government provided both financial and technical support to set up the Metal Workings and Machinery Industry Development Institute (MIDI) in order to foster the technical skills of SMEs in the metal and machinery industry. Then, in 1991, a team of technical experts from Japan International Cooperation Assistance (JICA) prepared the first detailed study of the Thai supporting industry, which became the basis for Japanese overseas development assistance. In 1995, the second JICA team prepared another comprehensive study of the Thai supporting industry, after which a large number of loans were given to the Thai government to upgrade the vocational colleges for teaching more advanced technical courses using new, sophisticated machinery. Meanwhile, the Thai government also launched some programmes to promote supporting industries. They included the Board of Investment's BUILD programme (or BOI Unit for Industrial Linkage Development) and the expansion of vocational courses on the use of new technology, such as computer-aided design (CAD), computer-aided manufacturing (CAM) and computer numerical control (CNC).

Therefore, technology transfers in the Thai automotive industry are influenced by such factors as: (1) the relationship between the MNEs and their local affiliates; (2) the technological capability of local part suppliers; (3) the export activities of local part suppliers; and (4) Japanese government assistance, as well as Thai government policy. However, the transfer of technology in the Thai auto industry still remains relatively limited to the production process for two main reasons: the strategic decision making and R&D activities have remained centralized at the automakers' headquarters, and the auto parts industry has been overprotected. Local affiliates find R&D activities unnecessary at the local level, since all decisions are made at the headquarters. Moreover, local OEM partmakers, which are joint ventures with the Japanese parts manufacturers, are not allowed to export their OEM parts. Even though government protection helped promote growth in the auto parts industry, such protectionism has somehow destroyed the endeavour for innovation and R&D in the local parts manufacturers. With protectionism in place, local auto parts manufacturers are assured of certain buyers; hence, the only thing the supplier has to do is follow product specifications.[1]

Employment in the automotive industry: from expansion to economic crisis

During the past three decades, the Thai automotive industry has been experiencing boom-and-bust cycles. Car production increased sharply from only 14,800 units in 1971 to 559,428 units in 1996, then dropped sharply to 158,130 units in 1998, and gradually picked up after that. The 1997–1998 downturn was not the first, but was the most severe. The second oil shock also caused stagnation in car production in the early 1980s (see Figure 1.3).

How does employment in the automotive industry respond to production fluctuations in the car market? In related issues, how does the skill composition of workers change in response to market expansion and recession, how does the industry acquire skilled labour during periods of rapid growth and how does it maintain those skills in the recession years?

Labour shortage and changes in employment structure in the early 1990s

Since the automotive industry is relatively intensive in terms of skill and technology, it tends to employ a relatively higher ratio of educated workers. Moreover, as the industry expanded rapidly from 1986 to 1996, the overall employment structure had moved towards employing workers with higher educational qualifications. Tables 9.1 and 9.2 indicate the constant decrease in the number of workers with only a primary education, and the gradual increase in those with secondary, technical, vocational and bachelor

Table 9.1 Employment structure according to educational level in the Thai automotive industry

Educational level	Average 1990–1996	1997
Primary to higher secondary	79	72
Technical or vocational training	12	19
Bachelor degree	5	6
Engineer	n.a.	3
Total	100	100

Source: TDRI, *Manpower Issues in the Thai Industries* (1998)

Note
The figures, which are percentages, show the sum of all workers in the auto assembly and the auto part sample firms

Table 9.2 Number of workers in the Thai automobile industry, classified by educational institution attended

Year	1986–87	1991–92	1996–97
Primary level or lower	71.7	67.9	61.7
Lower secondary	11.1	14.9	18.7
Upper secondary	4.1	4.2	5.5
Technical or vocational training	6.9	7.3	7.4
University/teacher college	5.0	5.5	6.5
Others	1.1	0.0	0.0
Total (percentage)	100.0	100.0	100.0
Total (thousands of persons)	195.0	314.4	441.2

Source: National Statistics Office, *Labor Force Survey*, August Round

degree levels. A Thailand Development Research Institute (TDRI) manpower survey for the automobile industry shows that the number of workers with primary-level certificates, accounting for 79 per cent of the sample firms' total employment during 1990–1996, declined to 72 per cent in 1997. Over the same period, the employment of technical workers, those with bachelor's degrees and engineers generally increased. During 1990–1996, the proportion of technical and vocational and university graduates with bachelor's degrees accounted for only 12 and 5 per cent, respectively. By 1997, these figures had risen to 19 and 6 per cent, respectively. In addition, engineers comprised 3 per cent of total employment in the Thai automobile industry.

A Labor Force Survey by the National Statistics Office also indicates a decrease in the number of workers having only a primary education, with corresponding increases in the hiring of workers with secondary, technical/vocational and university levels (Table 9.2).

Still, the employment of technicians, as well as engineers, in Thailand faces a few constraints, the first and foremost being the shortage of skilled technicians and engineering graduates. Employment in the automotive industry jumped sharply in the early 1990s in response to the rapid growth of the vehicle market that began in the late 1980s (see Figures 1.3 and 9.1).

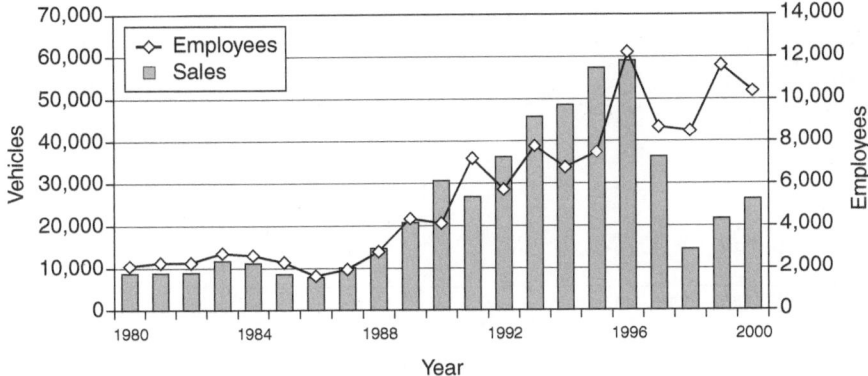

Figure 9.1 Employment and vehicle sales in the Thai automotive industry, 1980–2000

Source: Toyota Motor Thailand Co., Ltd, and National Statistical Office, *Labor Force Survey* (August)

Between 1990 and 1996, employment increased 33 per cent per year from about 41,000 workers to 122,300 workers. In the same period, vehicle production increased by only 13.9 per cent per year, implying that employment elasticity has a factor of almost 2.4. Such a sudden jump in labour demand resulted in a severe labour shortage. The automotive manufacturers, especially the new entrants, began to poach labour from the existing firms, resulting in a higher labour turnover. To protect their investments in training, most firms were forced to increase wages for their workers, particularly for skilled workers and engineers.[2] For example, during the early 1990s Toyota Thailand lost more than half of its trained engineers and a substantial number of its trained technicians in less than two years. Therefore, it decided to increase salaries of engineers several times in a single year in order to bring down the company's turnover rate. The salaries of engineers were more than doubled (Poapongsakorn *et al.* 1992).

At the same time, Japanese MNEs in both the automotive and the electronic industries had successfully lobbied their government to provide technical assistance and loans to the Thai government in order to alleviate the skill shortage.

In response to the shortage of personnel skilled in science and technology, the Thai government also moved quickly to expand the enrolments of science and technology students. Private vocational colleges and private universities expanded their enrolments not only in the areas of technical vocational education, science and engineering (see Figure 9.2), but also in commerce and business administration. By the mid-1990s, the manpower situation had improved. The number of engineering graduates increased substantially, particularly graduates from private colleges (see Figure 9.2).

Second, the rapid employment expansion also had an adverse impact on labour quality. During the period 1992–1995, the quality of Thai engineers declined markedly because technicians and vocational graduates were

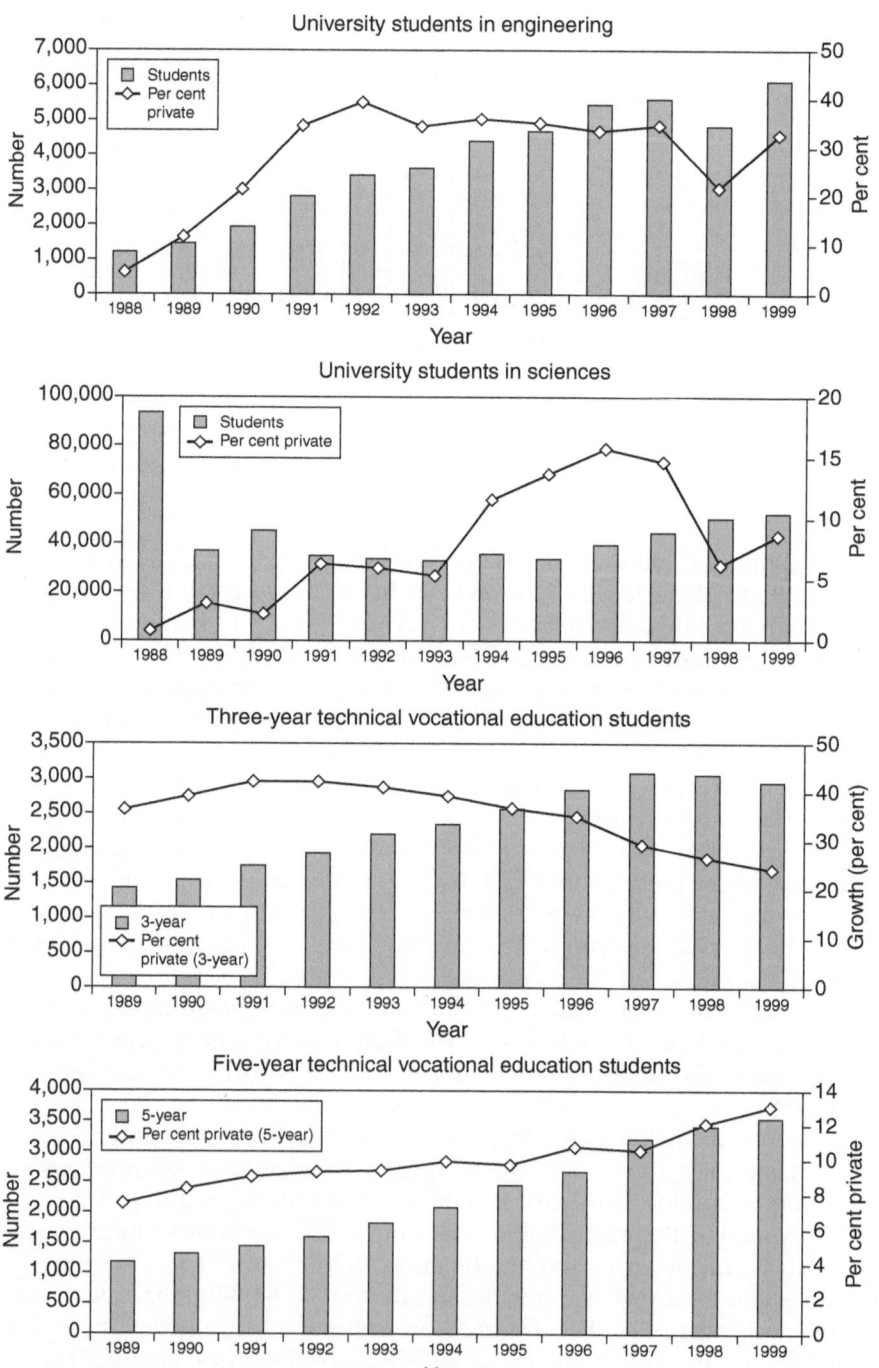

Figure 9.2 Number of students and graduates in Thailand

Source: Office of the Private Education Commission, Department of Vocational Education and the Ministry of University Affairs

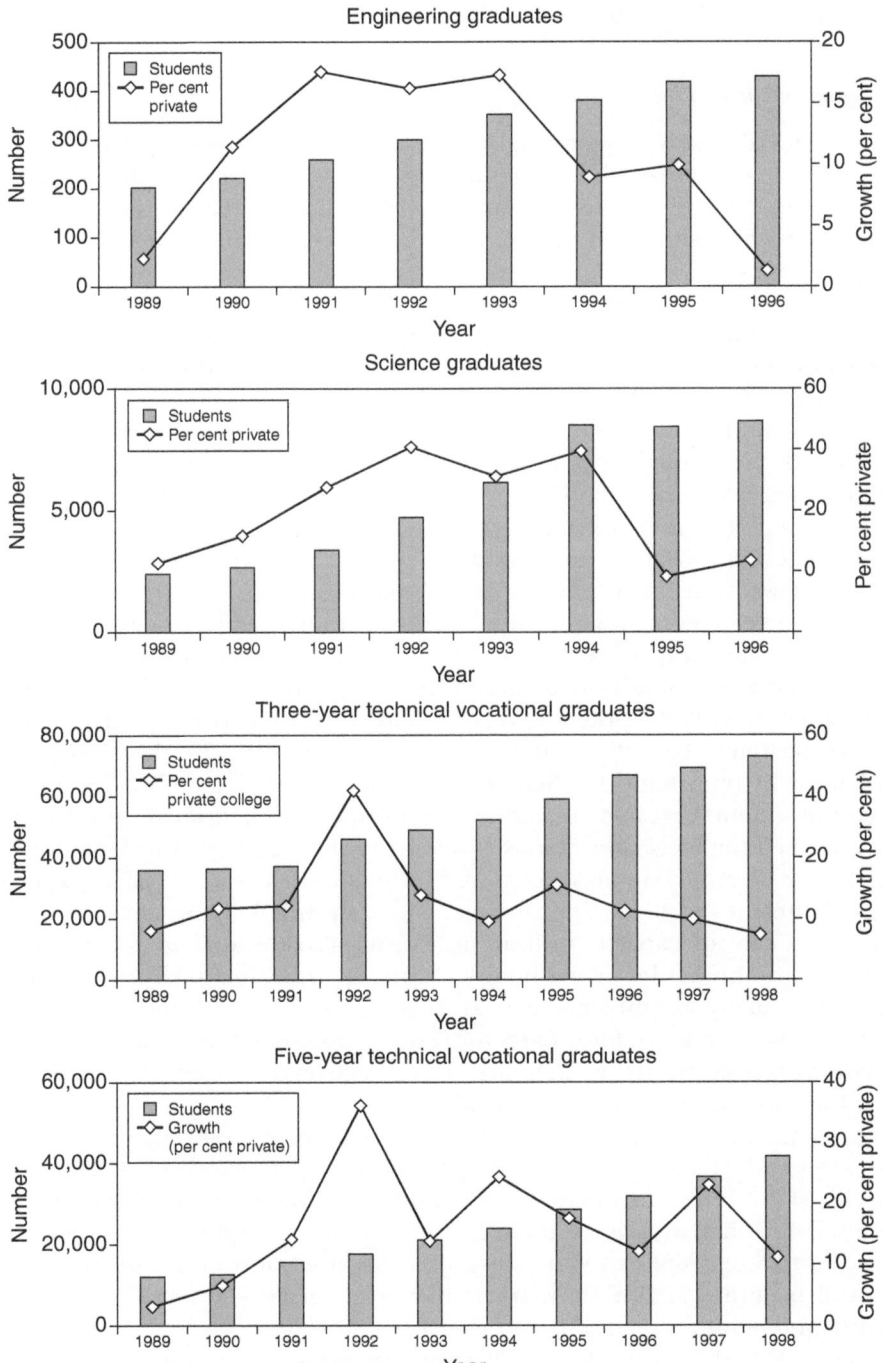

allowed to upgrade to engineer level. There were also newspaper reports that some manufacturers reduced their investments in training.

Economic crisis and labour hoarding

The economic crisis triggered by the flotation of the Thai baht on 2 July 1997 had severe impacts on the industry. Vehicle sales dropped sharply from 589,126 units in 1996 to 363,156 units in 1997 and 144,165 units in 1998. As a result, several automotive firms went bankrupt, causing a sharp drop in employment – not to be compared, however, with the decline in vehicle production. While employment declined by only 30.6 per cent in the same period, car production dropped by almost 72 per cent between 1996 and 1998 (see Figure 9.2). Obviously, automotive firms adopted labour-hoarding practices.

Skills acquired by automotive workers are specific to the industry (see below). If the manufacturers expect only a temporary downturn in their sales, it is not rational to lay off workers with specific skills because, when the industry recovers, they will not be able to hire back as many skilled workers as they need. It takes time and is costly to train new, inexperienced workers.

Interviews with medium-scale parts manufacturers confirm that they decided to keep most of their skilled workers and trained engineers, yet most companies were forced to reduce employment by laying off unskilled and semi-skilled workers, as well as those with general skills. However, keeping a large pool of skilled workers is very expensive and could easily drive firms into bankruptcy. To mitigate the impact of recession shocks, the MNEs had to lend a helping hand to their local affiliates in Thailand. Besides giving financial support and injecting additional capital into their local subsidiaries, they allowed the local subsidiaries to export vehicles and parts to Japan or to third countries (Poapongsakorn and Wangdee 2000). Industry experts regard these exports as unprofitable, but they will help to maintain the minimum rate of capacity utilization. Toyota also decided to relocate the assembly of one model of the luxury Camry car from Australia to Thailand. Moreover, many assemblers, such as Toyota, Honda, Ford and Mitsubishi, have plans to use more local parts for vehicles assembled in Thailand. These decisions can be viewed as measures to protect their investment in specific training and capital intensive production capacity.

Unfortunately, a number of SMEs went bankrupt, resulting in a sharp drop in employment. Many of those laid off were workers who had acquired specific skills. It is unfortunate for the industry to lose some of its investment in human capital, as most of them had to find other employment opportunities. Fortunately, the loss is not very great because most of the manufacturers still believe that ASEAN markets have high growth potential, and thus maintain relatively large numbers of workers.

Since 1999, the car market in Thailand has improved significantly. Vehicle sales increased from 144,065 units in 1998 to 262,189 units in 2000. Employment surged by 36 per cent between 1998 and 1999, but then dropped slightly in 2000. The employment increase in 1998–1999 was an overshooting response to the expected growth of domestic vehicle sales and surge in exports of both vehicles and parts. After realizing that most Thai companies still have serious debt problems and that future prospects of the global economy were not as expected, manufacturers began to revise their business plans. It should also be noted that, according to interviews with some auto-motive manufacturers, most new workers hired during 1999–2001 were semi-skilled contract workers who could easily be laid off when necessary.

Training and skill formation

Given the fact that the automotive industry is skill intensive, while most Thai workers have only primary education, it is necessary for manufacturers to provide intensive training for their workers. Broadly speaking, there are two approaches to training in the automotive industry, including both a simple method of on-the-job training (OJT) and off-the-job training (Off-JT), which may take place at the firm, at training centres, or at the parent company abroad. Many studies suggest that most training programmes in the Thai automotive industry tend to follow the Japanese pattern, which is highly intensive and systematic depending on the workers' education and seniority.

The main issues in this section are the processes of skill formation and upgrading, and their impact on the development of the automotive industry. The analysis will assess the external benefits of training and identify import-ant training problems in the industry. Two questions will be addressed: why do Japanese manufacturers have to make painstaking efforts to transplant the Japanese employment system and to invest heavily in specific skill train-ing, and is there any evidence of skill upgrading among Thai workers?

The Japanese employment and training system

The traditional Japanese employment system consists of four elements: heavy investment in specific training, lifetime employment, seniority-based pay and the enterprise union.

To increase workers' productivity, an employer had to provide intensive training and seek cooperation from the union so that conflict between workers and management could be avoided. Moreover, by treating its employees as valued members of an extended family, the management coaxed maximum effort and commitment from them (Hatch and Yamamura 1996: 147). However, in times of labour shortage, trained workers could be poached by other employers. The original employer would have to find means to reduce the incidence of labour turnover, the first of which would be to provide specific training. Under such circumstances, workers would have no incentive to leave the firm because their acquired skills would be useless in other firms. Even so, if workers do resign, the firm loses its invest-ment, so the firm has to provide enough of an incentive for these workers to stay on. The second way to inhibit labour turnover is to reward workers with a system of seniority-based wages in combination with a slow pace of promo-tion and lifetime employment. The seniority-based pay will ensure that trained workers will be partially rewarded for their higher productivity. The seniority-based pay and slow pace of promotion are also suitable for the Japanese firm, in which much decision making is decentralized and based on group consensus. In such an environment, it is difficult to measure the performance of each individual worker. In addition, the lifetime employment system in the Japanese firm is a double guarantee that the worker will not be laid off before he or she can reap enough benefits from the training received.

Why is most training in Japanese firms oriented towards a specific skill? In a firm in which most decisions are based on group consensus, the workers must be trained and socialized in such a way that they can fully understand each other and can jointly make effective decisions. Such interpersonal knowledge is highly specific because it cannot be exploited elsewhere. Moreover, the firm must share more firm-specific information with its workers if they are to make good and correct decisions together. Various kinds of group activities are also employed so that workers can help solve problems at every step of the process. Finally, the Japanese firm must also cultivate a sense of family among workers so they will cooperate as a group and accept the seniority system.

When Japanese investors began to establish plants in Asia, they had to modify the employment and training systems to suit the local conditions. Forty years ago, when Japanese enterprises first invested in Asia, most Asian workers were agricultural migrants who were not used to industrial discipline. The Japanese firms had to instil work habits and nurture company loyalty. Another major difference between the Japanese and the South-East Asian economies is that most workers in the latter countries are less educated. Because of these differences, the Japanese firms have to rely more on OJT and general training. For the less educated South-East Asian workers who do not understand the Japanese language, exposure to more intensive OJT is necessary. Managers spend marathon sessions providing detailed verbal instruction to employees, rather than relying on written manuals. Unlike Japanese workers, who have the opportunity to rotate their jobs, Asian workers are trained narrowly, to perform only specific tasks. Our survey of the Japanese firms in Thailand finds that most of them do not have the job rotation policy. The Japanese firms in Thailand and other South-East Asian countries also provide or sponsor extensive general training for local workers. The training, which is mostly in a form of Off-JT, includes courses on safety in the workplace, instrumental and machine calibration, quality circles and management. This kind of general training directly increases the productivity of firms in both automotive and non-automotive industries.

Other modifications in the Japanese employment system include a less rigid adherence to lifetime employment and less dependence on bonus payments as a cornerstone of the performance-based pay system. As mentioned above, when there was a labour shortage in the late 1980s, many Japanese firms in Thailand were forced to increase the salary of skilled workers and adopt a faster pace of promotion. Finally, most of the Japanese firms in Thailand do not encourage union activities, let alone union formation.

Pattern of training and skill upgrading

As mentioned above, the automotive industry is relatively more skill and technology intensive than other industries, except the electronics and computer industries. Automotive firms therefore have to recruit a relatively higher proportion of educated workers. The minimum educational requirement for workers recruited by large and medium-scale automotive firms is lower secondary education, compared to primary education level for other industries. The educational composition of industrial workers as shown in Table 9.3 confirms that the automotive industry hires a larger proportion of educated workers. Moreover, the automotive industry also provides more frequent training for its workers. Table 9.4 presents information on the amount of training based on two separate surveys of industrial workers, one in 1991 and the other in 2001. Although the two surveys differ in the sample size and coverage, they clearly confirm that there is a greater amount of training provided in the automotive industry than in the other industries covered by the survey. Interviews with two large auto assemblers, namely Honda and Toyota, also indicated that their formal HRD programmes are more intensive than the industry average. They have institutionalized training programmes by providing formal essential training courses for skilled workers and technicians in both the car assembling and the auto parts industry. The concept of their HRD project is similar to McDonald's Hamburger University, whereby the company attempts to create the College of Automotive Industry.

But there is one exception. Automotive workers receive less OJT than those in other industries in terms of both the percentage share of workers receiving OJT and the length of OJT per worker. This information contradicts what has been said above. One explanation may be that the automotive workers have a higher minimum education level and therefore do not need so much OJT as workers in other industries.[3] Not only does the automotive industry provide more intensive training than other industries, but its

Table 9.3 Educational composition of workers in Thailand, 2001

Education level	Auto parts		Other industries	
	No. of workers	*%*	*No. of workers*	*%*
Primary	18	3.8	198	14.2
Lower secondary	121	25.8	249	17.8
Upper secondary	171	36.5	480	34.3
Academic	99	21.1	361	25.8
Vocational	72	15.6	119	8.5
Vocational diploma	99	21.1	216	15.5
University	60	12.8	255	18.2
Engineering and sciences	40	8.5	182	13.0
Others	20	4.3	73	5.2
Total	469	100	1,398	100

Source: TDRI, *A Survey of Education and Training* (August 2001)

Table 9.4 Training incidence in various Thai industries

Industry	1991		2001	
	Days/ worker-year	Times/ worker-year	Days/ worker-year	Times/ worker-year
Auto parts				
All formal training	178.0	3.3		
In-house training	2.4	n.a.	3.33	4.53
Training centres	3.4	n.a.	n.a.	1.74
Parent firm abroad	10.7	n.a.	n.a.	1.53
Auto assembly				
All formal training	20.9	4.0		
In-house training	4.5	n.a.	4.11	1.86
Training centres	7.3	n.a.	n.a.	1.00
Parent firm	9.8	n.a.	n.a.	1.40
Electronics				
All formal training	18.6	2.2		
In-house training	4.2		3.15	4.25
Training centres	4.3		n.a.	1.65
Parent firm	10.1		n.a.	1.38
Food processing				
All formal training	0.0	0.0		
In-house training	0.0	0.0	2.62	4.21
Training centres	0.0	0.0	n.a.	1.81
Parent firm	0.0	0.0	n.a.	1.65

Sources: 1991: Poapongsakorn *et al* (1992: 32); 2001: TDRI, *A Survey of Education and Training* (August 2001)

Note
Electronics in 2001 includes (a) semi-conductor and (b) hard disk

pattern of training also shows an interesting process of skill upgrading in both production and management technology.

If one examines the pattern of labour training in most large and medium-scale factories, it is difficult to distinguish differences between industries. In a world where information flows rapidly, the training technology invented by one industry is likely to spread quickly and be adopted by other industries. Most, if not all, have similar training programmes for all levels of employees. Newly recruited workers have to undergo a one- or two-day orientation programme before being sent to the shop floor, where they are trained to perform their tasks by a foreman or senior employee. In some rare cases, they may have to attend a few days of formal training courses. The immediate training objective of OJT is to teach newly hired production workers some necessary basic skills and knowledge to perform their jobs. The skilled workers also have to undergo OJT every time there is a change in the production process, or a new product or new machinery is introduced. More importantly, on the middle or supervisory levels, employees attend formal training programmes organized by the company, by the machinery suppliers, as well as by independent training centres. The training programmes are either production- or management-related courses, depending upon the needs of the companies and the employees' responsibilities. At home, senior engineers or senior management conduct training programmes. In addition, some skilled employees in local subsidiaries are sent to attend training programmes at the company headquarters overseas, and after their return they arrange short courses for other employees.

The pattern of training in the automotive industry over the past decade, however, has shown some interesting characteristics. Two examples concerning training in die casting and in management will sufficiently illustrate an emerging process of skill upgrading in Thailand.

In the 1970s, when Siam Navaloha – a subsidiary of the Siam Cement Group – decided to enter the auto parts industry, it had to recruit experienced workers with only primary education from small metal-casting firms. Experienced workers were assigned to train newly recruited young workers with three years of vocational education, using the OJT approach. At the time, it took eight to ten years for an average apprentice (with primary education) in a small-scale metal-casting firm to master the skill of metal die casting. But thanks to the systematic training programme in Siam Navaloha, it took only eighteen to twenty-four months for a young three-year vocational graduate to learn the skills of metal die casting (Poapongsakorn *et al.* 1992). There are at least two reasons why workers in the modern factory would master the skills faster than their counterparts in the small factory, even had both been trained by the same experienced worker. First, the more educated worker in the modern factory is more trainable than the less educated one. With some basic drawing background, for instance, vocational graduates could quickly understand the process of metal die casting. There is also evidence from Siam Navaloha and other metal firms that workers with high school certificates in an academic stream were more easily trained to use modern machinery than those with vocational education because the former had a stronger mathematical background (ibid.). Second, many modern factories also adopt more systematic training methods. Workers are taught to do their jobs step by step. For example, they have to know the basic properties of metal and to be able to draft a drawing of the object to be produced. By contrast, the apprentices in the small factory have to observe and learn processes almost entirely by themselves.

Ten years ago, Japanese firms in Thailand began to use CAD/CAM in place of real clay models for the design of metal parts. The CNC machines were also used in the production of metal dies and moulds. Although computerized machines could substantially increase productivity of die production, the existing workers had to be retrained and more highly educated workers had to be hired. At the time, only Japanese engineers and technicians had the knowledge to use the new computerized machines. According to interviews with veteran engineers in the die factories, it took more than a year for a technician with five years of vocational education after lower secondary school to develop the skills necessary to use these machines. But today it takes only three to five months to train new workers. Young school leavers have already been exposed to the new technology in their schools.

But, more importantly, after a one- to two-month training they will be working with a group of experienced workers who can teach them how to solve the more practical problems in their work.

There are at least four reasons why it takes a shorter time and is less costly to train the new generation of workers. First, the existence of a pool of experienced workers generates spillover effects on the training of new workers. Second, there may be an increasing return to training since it has high fixed costs. Third, there may be economies from learning, i.e. the firms' average costs in the long run may decline because as the cumulative output produced increases, workers absorb new technological information and become more experienced at their jobs. As a result, the average training cost declines with the increasing cumulative output. Finally, the new generation of workers with a more mathematical background is more susceptible to training than the older generation.

With regard to skill upgrading in management training, throughout the past twenty-five years Thai workers have been heavily exposed to Japanese-style management training, ranging from early-morning physical exercise and 5S to *kaizen* and TQM practices. As a consequence, the Thai automotive industry has enjoyed higher productivity from improved management practices. Since the mid-1990s, when partmakers began to increase their exports to Europe, they have had to adopt the ISO standards, which require firms to document all work procedures. Our interviews show that the ISO practices may have a positive impact on productivity. Some factory managers reported that the detailed documentation of work procedures may have reduced the accident and injury rates in some production areas because the workers are now properly trained and have to do their work according to the steps written in the work manual. Moreover, human resource managers of a few companies have begun to formalize the OJT for most production activities, resulting in a more systematic method of training. Likewise, introduction of the QS 9000 standards in firms that want to do business with American carmakers or first-tier part suppliers has some positive impact on training and consequent effects on productivity.

In summary, the process of skill upgrading in the Thai automotive industry has been influenced by several factors, particularly the introduction of new technology (embodied in CAD/CAM and the CNC machines), a drive for exports and the requirements of the new business partners, such as European customers and American carmakers. There are also other important agents helping the process of skill formation, perhaps the most important of which is the Japanese government, which has played a significant role in helping the Japanese automobile and electronics MNEs to establish regional production networks in Asia (Hatch and Yamamura 1996). The Japanese government has provided both technical and financial assistance to Thailand to develop supporting industry and SMEs. The technical assistance included two assessment rounds of the technological capability of the automotive and electronics industry and the SMEs. The Japanese government also provided the financial assistance that helped to found the MIDI in the 1980s and to establish the privately run Thai–Japanese Technology Promotion

Association (TPA), technical assistance for the semi-government, the Technology Improvement Institute (TII), and loans for the government to build Provincial Skill Development Centres. These training centres have provided extensive training in both technical and management areas for several thousand workers over the past twenty years.[4] Moreover, the Japan Bank for International Cooperation (JBIC) and, in the past, the Overseas Economic Cooperation Fund of Japan (OECF) have given long-term, low interest loans to the Thai government to modernize the training and teaching equipment in public vocational colleges. Thousands of vocational teachers have also been trained or educated in Japan. Without such assistance, it would be almost impossible for the Thai automotive industry to train the large and increasing number of workers it needs. The availability of a large pool of skilled workers and vocational education graduates has obviously played a significant role in attracting American automakers to invest in Thailand in the late 1990s.

The German government has also played an important role in providing technical assistance in the area of industrial skill development. In the 1970s, it helped establish a technical vocational school to produce graduates with necessary industrial skills. The school was later upgraded to become the King Mongkut Institute of Technology. In 1995, Germany also helped the Thai government to establish the Thai–German Institute, the main objective of which is to provide a broad range of services aimed at upgrading skills and practices related to automation, CAD, CNC and precision machining of tools, dies and moulds. Between 1998 and 1999, it provided training services for 2,720 workers from 656 firms. Most customers are small-scale enterprises (Nexus Associates 2000: vol. 1, 38–43).

The training centres have also helped reduce the cost of training, particularly in SMEs. The important training centres are TPA, the Institute of Productivity Improvement, and several public training centres that offer both technical and management courses. Since there are thousands of part suppliers in the automotive, electrical and electronics industries, these training centres can provide a large number of training courses at low tuition fees. Many training courses are subject to economies of scale because they involve large fixed costs. As a result, automotive employers, including SMEs, have been more willing to send their workers to be trained at these centres.

Notable too is the role of some local part suppliers, particularly the Siam Cement Group, the Sammitr Group, the Somboon Group, the Summit Auto Parts Group and the Thai Rung Union Group. Thanks to their vision of the industry and their entrepreneurial spirit, they were the leading local firms in the automotive industry until the economic crisis broke out in 1997. The Siam Cement Group played the major role in lobbying the government to adopt the local content policy in the 1970s. But more importantly, these companies, particularly the Siam Cement Group, had been the training ground for engineers and vocational graduates who later moved to other companies. They were the pioneers who provided systematic and modern methods of training for their workers. A 1991 study of labour training found that their training programmes enabled new workers to master basic skills in a shorter time than their counterparts in smaller firms (Poapongsakorn *et al.* 1992).

Training problems

There are still some training problems in the Thai automotive industry. On the employees' side, there are two major problems. First, most workers with diplomas in vocational education do not possess the strong background in basic sciences that is necessary for them to understand the production techniques or quality of materials used in the production process. Second, many of these technicians and employees have problems with their communication abilities, particularly with regard to writing. Thai engineers also have problems communicating in English with their Japanese employers, neither of whom, as a rule, are good at speaking English. Our interviews with both Thai and Japanese engineers who work in the same department revealed large information and knowledge gaps concerning production technology and technical aspects of the production process. Such language deficiencies not only affect the degree of technology absorption and ability to master the training received fully, but also explain why it is difficult for Thai engineers to be promoted to the top management level, where managers have to communicate in Japanese.

On the employers' side, there is less incentive for the Japanese firm to transfer more technology-intensive products or R&D activities to its foreign partners. There are at least three reasons. First, firms outside Japan may still have low levels of technological capability. Second, the recent division of labour between automotive MNEs and suppliers resulted in R&D activities being carried out by the former. Third, Japanese partners prohibited their foreign part suppliers, by contract, from exporting their products. For these reasons, training at the local partmaker firms in Thailand is concentrated only in the area of production technology and management.

Effects of training and education

The preceding analysis shows how the workforce in the automotive industry has acquired its skills, which in turn enhanced manufacturers' productivity. But trained workers need to be rewarded for their higher marginal productivity, otherwise they do not have any incentive to stay with the company. What, then, is the impact of training on workers' wages? Do education and training play complementary roles in explaining the workers' wages, and what are the factors that influence decisions as to which workers are to receive training?

This section analyses the impact of training and education, using a survey of 1,867 workers from twenty firms in four industries, namely, the auto parts, the semiconductor, the disk drive and the food processing industries. Two separate sets of estimates are compared: the estimates for the auto parts industry vis-à-vis those for all industries. The survey of the auto parts workers comes from two firms producing vehicle air compressors and two firms in the mould- and die-casting industry. The sample firms in each industry produce very similar kinds of products, using more or less the same technology. The survey was carried out between June and August 2001.

The data set used in this study is a panel of data. In each firm, about 50–100 production workers were interviewed face to face. Each worker was asked to provide information on earnings and training for the 1998–2001 period, and was also asked about previous work history and family background. The researchers also interviewed the corresponding employers using a structured questionnaire that contained questions on financial statements, output and employment, training, turnover, wages and fringe benefits, hours of work and changes in technology in the past five years. But the employer survey is not yet complete. The results in this chapter are still very preliminary since additional information from eight more firms has yet to be included and the data have not yet been satisfactorily tidied up.

Workers' characteristics and training: a comparison

Before I present the results, it is useful to describe some important characteristics of the sampling of workers. The data in Table 9.3 confirm that lower secondary education is the minimum requirement for workers in the auto parts industry, since 96 per cent of them have at least lower secondary education compared with 86 per cent in other industries. The table also shows that the auto parts industry hires a larger proportion of vocational graduates than other industries, i.e. almost 37 per cent compared with 24 per cent, because the auto parts industry is more skill intensive. In terms of workers' ages (Table 9.5), the auto parts industry is not very different from other industries.

Table 9.5 Number of workers by age and length of work experience (%)

Age/experience	Auto parts	Other industries
Age (years)		
< 20 years	1.9	5.7
20–24	24.5	20.5
25–29	37.1	36.3
30–39	31.8	32.4
40–49	3.8	4.2
50 and over	0.9	0.9
Experience (years)		
≤ 2	14.3	14.6
3–5	18.8	19.0
6–10	34.1	36.7
11–15	24.5	21.5
16–20	4.9	5.4
> 20	3.4	2.8
Total %	100	100
No. of workers	469	1,398

Source: TDRI, *A Survey of Education and Training* (August 2001)

However, it has slightly higher percentages of workers both with less than two years and with more than ten years of work experience. The higher proportion of the former category may reflect the fact that the rapid recovery of auto sales since 1999 has encouraged firms to hire more new workers, some of whom may be in their mid-20s. That the auto parts industry has a slightly higher proportion of workers with more than ten years of experience also shows that work experience is quite important for the industry, which has been dominated by the Japanese seniority-based employment system.

The information on job position in Table 9.6 is also consistent with the data on worker education levels. The auto parts industry has a higher percentage of technicians than other industries. Most technicians have vocational education. But the auto parts industry has a much smaller share of team leaders and supervisors. This may explain why other industries tend to hire more workers with upper secondary academic education or those with university degrees, who can be more easily trained into foremen and supervisors. That there is a smaller percentage of foremen and supervisors in the auto parts industry may be because the industry is characterized by its small scale of production, and the nature of the assembly process does not require so many small groups of workers working together as in the other industries.

Perhaps the most interesting information concerns the training. Although the auto parts industry is more skill intensive than other industries, it provides less OJT in all aspects of measurement. Table 9.7 shows that OJT in the auto parts industry is received by a smaller proportion of workers and occurs less frequently, and for a shorter period, but the auto parts industry does provide more in-house Off-JT than the other industries. This may be because it hires a larger proportion of vocational education graduates as production workers. These workers gained some practical work experience in school, but they need to be given more formal in-house training. The industry also sends larger proportions of workers to be trained at the independent training centres. But the auto parts and the other industries do not differ in terms of their sending workers abroad to be trained by their parent firms.

Table 9.6 Job positions of workers, 2001

	Auto parts		Other industries	
	No. of workers	%	No. of workers	%
Production worker	315	67.2	829	59.3
Team leader or assistant	26	5.5	123	8.8
Supervisor/foreman or assistant	28	6.0	145	10.4
Head of section/head of department	26	5.5	78	5.6
Technician	50	10.7	122	8.7
Engineer	13	2.8	87	6.2
Manager	11	2.4	14	1.0
Total	469	100	1,398	100

Source: TDRI, *A Survey of Education and Training* (August 2001)

Table 9.7 Training incidence by type of training

Training incidence	1998		1999		2000		2001	
	Auto parts	Others	Auto parts	Others	Auto parts	Others	Auto parts	Others
1 No. of workers	308	1,024	367	1,125	434	1,297	469	1,398
2 No. of trained workers (%)								
OJT	55.8	67.6	53.1	66.7	56.9	66.5	55.4	63.0
Off-JT	62.3	70.0	58.9	72.2	68.2	78.2	67.0	62.3
Training centres	9.7	9.4	12.5	11.3	14.8	10.3	9.4	6.4
Parent firms	5.2	6.9	6.3	7.4	11.5	9.6	8.3	5.6
3 No. of times trained per worker								
OJT	4.0	5.4	3.8	5.1	4.7	5.3	3.7	4.7
Off-JT	3.8	3.1	4.2	3.1	3.6	3.0	3.4	2.7
Training centres	1.4	1.7	1.7	1.6	1.5	1.7	1.6	1.7
Parent firms	1.3	1.6	1.1	1.6	1.7	1.7	1.5	1.5
4 Length of training (days/workers)								
OJT	0.6	1.3	0.6	1.2	1.0	1.4	0.6	1.2
Off-JT	3.8	3.2	6.5	3.8	3.6	3.1	3.4	2.7
Training centre	—	—	—	—	—	—	3.9	5.2
Parent firms	—	—	—	—	—	—	18.0	10.3

Source: TDRI, *A Survey of Education and Training* (August 2001)

Note

OJT: on-the-job training; Off-JT: off-the-job training

There are some major differences between the training courses given in the auto parts and the other industries. The auto parts firms tend to provide more Off-JT in the areas of safety regulations and ISO/QS 9000 than other industries do (Table 9.8). But when the firms send workers to the training centres, those in the other industries prefer to send workers to attend the ISO programme. This is because the ISO programmes can be applied to all industries, while the QS 9000 programme is more industry specific and thus has to be provided in-house. At the training centre, all manufacturing firms send most of their workers to attend courses on production technology. But

Table 9.8 Number of workers receiving off-the-job training and attending training at a training centre in 2001, listed by courses

	Auto parts		Other industries	
	No. of workers	%	No. of workers	%
Off-JT				
Orientation for new employees	10	1.5	15	0.8
Daily tasks	117	17.3	373	20.5
Safety regulations and rules	198	29.2	448	24.7
Maintenance, operation and mechanics of machines and tools	61	9.0	134	7.4
Orientation for new production process	38	5.6	128	7.0
Orientation for new machines and equipment	32	4.7	76	4.2
Orientation for the production of new product	5	0.7	14	0.8
Improving skills in general	46	6.8	118	6.5
Finance and accounting	1	0.1	12	0.7
Marketing	1	0.1	14	0.8
Improving working efficiency and problem solving	33	4.9	104	5.7
Management and leadership	38	5.6	99	5.5
ISO, HACCP and labour standard, QS 9000	97	14.3	281	15.5
Total	677	100	1,816	100
Training centre				
Marketing	1	0.9		
Industrial standard, i.e. ISO	13	11.7	49	20.7
Management	14	12.6	39	16.5
Law and government regulations			6	2.5
Economy			1	0.4
Production technology	48	43.2	69	29.1
Measurement	1	0.9	7	3.0
Workplace safety	14	12.6	25	10.5
General information	15	13.5	19	8.0
Cannot remember	5	4.5	22	9.3
Total	111	100	237	100

Source: TDRI, *A Survey of Education and Training* (August 2001)

the percentage of auto parts workers who attend the courses on production technology is much higher than that for the other industries, i.e. 43.2 per cent versus 29.1 per cent for other industries.

The last interesting pattern concerning training is that OJT and Off-JT tend to be positively correlated. Table 9.9 shows that in 2001, most of those who received OJT also received Off-JT (70 per cent), and vice versa, but with smaller percentage shares (62.8 per cent). This implies that OJT and Off-JT are complementary.

Factors affecting training[5]

It is hypothesized that the probability that a worker receives training depends on years of education, age, work experience (or tenure), gender and job position. The squares of the first three variables are included to reflect the non-linear relation with training. The dummy variables representing the company job positions and the time periods are also included. The OJT and Off-JT regressions are jointly estimated by a bivariate probit regression, using STATA software. To correct the problems of heterocedasticity, the estimates use the companies as the cluster. The results are reported in the two tables designated Appendix 9.A1 and 9.A2 at the end of this chapter. There are two sets of estimates, i.e. those for all industries and those for the auto parts workers. The estimates are also done separately for all workers, production workers and non-production workers.

In the regressions of workers from all industries, the predictive powers of OJT regression and Off-JT regression for all workers and non-production workers are quite similar, and are low. In the OJT regression for all workers, only that for males is statistically significant, with a negative coefficient. This means that male workers are less likely to be trained, or even have less opportunity to receive OJT, than female workers, and have less impact on OJT. Nevertheless, for non-production workers, tenure and squared tenure are statistically significant in the OJT equation, with negative and positive coefficients, respectively. This means that, as the length of work experience increases, non-production workers will have less opportunity to receive OJT, until the threshold number of years of experience is reached. After that, they will have more opportunity to receive OJT.

On the other hand, no human capital variable is significant in the Off-JT regression for all workers and non-production workers. Therefore, the Off-JT regression for all industries has lower predictive power and does not provide a significant relationship among all explanatory variables in the equations.

But the results for the auto parts regressions contrast with one another, since the regression for both all workers and non-production workers in the auto parts equations have more predictive power than the regressions of workers for all industries. Meanwhile, the OJT regression for all workers in the auto parts industry is more predictive than OJT regression for non-production workers. Four important human capital variables are statistically significant in the OJT equation. They are: years of schooling, years of schooling squared, age squared and tenure squared. The negative coefficient of

Table 9.9 Number of auto parts workers receiving on- and off-the-job training, 1998–2001

| | | | OJT 2001 | | | | | |
| | | | No. of workers | | | % | | |
			Yes	No	Total	Yes	No	Total
Off-JT	No. of	Yes	784	401	1,185	66.16	33.84	100
2001	workers	No	357	325	682	52.35	47.65	100
		Total	1,141	726	1,867	61.11	38.89	100
	%	Yes	68.71	55.23	63.47			
		No	31.29	44.77	36.53			
		Total	100	100	100			
			OJT 2000					
			No. of workers			%		
			Yes	No	Total	Yes	No	Total
Off-JT	No. of	Yes	903	206	1109	81.42	18.58	100
2000	workers	No	407	351	758	53.69	46.31	100
		Total	1,310	557	1,867	70.17	29.83	100
	%	Yes	79.14	28.37	59.40			
		No	35.67	48.35	40.60			
		Total	100	100	100			
			OJT 1999					
			No. of workers			%		
			Yes	No	Total	Yes	No	Total
Off-JT	No. of	Yes	681	263	944	72.14	27.86	100
1999	workers	No	346	577	923	37.49	62.51	100
		Total	1,027	840	1,867	55.01	44.99	100
	%	Yes	59.68	36.23	50.56			
		No	30.32	79.48	49.44			
		Total	100	100	100			
			OJT 1998					
			No. of workers			%		
			Yes	No	Total	Yes	No	Total
Off-JT	No. of	Yes	610	253	863	70.68	29.32	100
1998	workers	No	298	706	1,004	29.68	70.32	100
		Total	908	959	1,867	48.63	51.37	100
	%	Yes	53.46	34.85	46.22			
		No	26.12	97.25	53.78			
		Total	100	100	100			

Source: TDRI, *A Survey of Education and Training* (August 2001)

Note
OJT, on-the-job training; Off-JT, off-the-job training

education in the OJT regression implies that if the worker has more years of education, he or she has less chance to receive OJT because there may already have been exposure to some degrees of training while the worker was studying in a vocational institution. Since the tenure coefficient is not significant, while the squared tenure is, one can interpret this as meaning that as tenure increases, the worker has a greater chance of being trained.

Aside from OJT regression, the results of Off-JT regression for all workers in the automobile industry have more predictive power than the results for non-production workers. As mentioned, age, work experiences and their squares and the male dummy are statistically significant. That is, the workers in the automobile industry will have an increasing probability of receiving Off-JT as their age and work experience increase. After that, they are less likely to receive Off-JT. Interestingly, the male dummy is significant in the Off-JT regression for all workers in the automobile industry. That is, the male worker is more likely to receive Off-JT than the female is.

Conclusion and some policy implications

This study shows that automotive firms in Thailand have used various means to cope with the problems of skill shortage and labour surplus, ranging from making use of market mechanisms, creating the employment system and training institutions, to lobbying the government to help solve the problems. During the 1980–1995 period of rapid growth in the automobile market, there was a serious shortage of skilled labour. The Japanese affiliates were forced to train workers themselves. To prevent workers from quitting the firms after their training, the Japanese companies invested in specific skill training by transplanting the Japanese employment system. The big Japanese automakers invested heavily in HRD, ranging from providing OJT and in-house Off-JT, sponsoring workers to attend the training centres, to sending thousands of workers to be trained in Japan. Yet these measures were not enough to prevent workers from being poached by other companies when vehicle sales jumped sharply in the late 1980s and early 1990s. Many Japanese carmakers, therefore, were forced to give their trained engineers and technicians several large salary increases.

Recognizing the weakness of the Thai educational system and the Thai government's slow policy response, the Japanese MNEs (in both the automotive and the electronics industry) had lobbied their own government to provide financial and technical assistance to the Thai government in order to produce more skilled labour. The Japanese government not only provided support and loans for public educational institutions, particularly the vocational colleges and educational institutions, but also helped establish non-profit training centres, such as the Thai–Japanese Technology Promotion Institute. Moreover, Thai private training institutes also responded to the labour shortage by quickly expanding enrolments in vocational education and engineering.

The efforts of the Japanese companies and their government, together with the efforts of a few Thai auto parts firms, have resulted in an increasing return to training and have generated spillover effects. Not only does the industry have access to the larger pool of skilled workers and highly trainable young workers, but the private costs of training to both the employers and the employees have also declined substantially in terms of the average training fees and average length of training period because increasingly more workers are trained. Another important source of cost reduction is a fall in the time required to train new workers and the higher labour productivity that is now available from the team of more highly skilled workers. The availability of the large pool of skilled workers, technicians and engineers not only was one of the critical factors that attracted new investment from the US and European carmakers in the mid-1990s and early 2000s, but also benefited employers in other industries, who did not have to invest in training.

The fact that most of the automotive workers' skills are specific helps to explain why the decline in the percentage of automotive employment was much smaller than the sharp drop in vehicle sales during the 1996–1998 period. Automotive producers had to protect both their investment in specific training and their capital-intensive production capacity. Soon after the crisis, both car assemblers and auto part suppliers shifted a significant proportion of production towards exports.

From a survey of 1,867 industrial workers from four industries, the preliminary econometric results show that neither the OJT nor the in-house Off-JT directly affect workers' earnings. However, in the auto parts industry, educated workers' wages will grow faster than the wages of those who are not educated. The results are valid for both production and non-production workers. However, the results of the interaction between work experience and the incidence of Off-JT received in the previous period are not significant, and the results are similar to those between all industries and the auto parts workers.

The human capital variables have significant impact only on the OJT of the auto parts workers, and not on their Off-JT, despite the fact that auto parts workers have a higher incidence of OJT than workers in other industries. This result may be consistent with an interpretation that OJT in the auto parts industry tends to be job specific.

The estimated equations of OJT for the auto parts workers are consistent with the Japanese employment system. Those non-production workers with more than twelve years of education and those with more than six years of experience in the same company will have a higher probability of receiving OJT. However, no human capital is statistically significant and has significant impact on Off-JT estimates for non-production workers.

The econometric results reported in this chapter are still preliminary, as the data have not yet been thoroughly processed, and more sample workers will be included in future work. But so far, the results are consistent with the Japanese employment pattern. Surprisingly, training will benefit only those workers with longer work experience or those with a higher level of education. If there is not enough incentive for workers, particularly the less educated workers on the production lines, then why do auto parts firms tend to provide more Off-JT? It is hypothesized that the training should have significant impact on the company's productivity. This hypothesis will be tested empirically after the employer questionnaires have been collected and processed.

This study argues that there are increasing returns to training services, as well as spillover effects from training. The argument provides a strong rationale for government subsidization of training. However, since training in itself does not directly enhance workers' earnings, and since the study shows that education and training are complementary, the most efficient government intervention is to invest in higher education. Such a policy will indirectly affect workers' welfare because the more highly educated workers are more likely to be trained by their employers.

Notes

1 In 2000, the highly protective local requirement regime was abolished and replaced by the moderate tariff rate of 27 per cent for the imported parts.
2 Several studies pointed to a serious shortage of manpower trained in science and technology in the late 1980s, e.g. Kritaykirana and Brimble (1988) and TDRI (1989).
3 Note that this explanation may not contradict the earlier assertion that the Japanese investors had to put more efforts into providing OJT for their workers, particularly during the early years of their investment in the 1970s and 1980s.
4 The TII was originally the Center for Productivity Improvement. It was one of the first autonomous institutes set up by the Ministry of Industry, but did not receive any assistance from the Japanese government until 1994, when it was upgraded into an independent institute. Its main objective is to enhance productivity in all economic sectors of the Thai economy. The main activities are to provide training and consulting services on issues relating to quality assurance, ISO 9000/14000 and productivity, to promote productivity awareness, and to conduct research relevant to productivity issues. Between October 1997 and September 1999, it provided consulting services to 256 companies, 78 per cent of which employed more than 100 employees. The number of customers has been growing steadily, with an average of thirty-two new customers added each quarter (Nexus Associates 2000: 61).
5 This section is based on a survey of 1,867 industrial workers in four industries, i.e., auto parts, food processing, electronics and integrated circuit. The survey, carried out in 2001, covered 20 companies. In each company 50–100 workers were interviewed using a structured questionnaire. Wage and training information between 1998 and 2001 were collected.

References and further reading

Abdulsoma, K. (1997) 'ASEAN Automobile Industry in the Emerging Economic Integration Environment: Corporate Strategies and Government Policies', paper prepared for the 2nd Export Meeting of the Interregional Project 'Transnational Corporations and Industrial Restructuring in Developing Countries' organized by the Division of Investment, Technology and Enterprise Development, UNCTAD, Geneva, November.

—— (2000) 'Government Policy, Liberalization and Globalization of the Automobile Industry in Thailand', *Business and Society, Interdisciplinary Journal*, 1 (2) (June).

Brooker Group (1997) *Automotive Industry Export Promotion Project. Thailand Industry Overview. Report*, prepared for the Office of Industrial Economics, Ministry of Industry, Royal Thai Government.

Findlay, C.C. and Abrenica, M.J.V. (2000) 'The ASEAN Automotive Industry: Challenges and Opportunities', PECC conference paper, August.

Fujita, K. and Hill, R.C. (1997) 'Auto Industrialization in Southeast Asia: National Strategies and Local Development', *ASEAN Economic Bulletin* 13 (3): 312–332.

Hatch, W. and Yamamura, K. (1996) *Asia in Japan's Embrace: Building a Regional Production Alliance*, Cambridge: Cambridge University Press.

Kato, S. (1992) 'Thailand's Auto Industry', *Asian Monthly Review*, October.

Mori, M. (1999) *New Trends in ASEAN Strategies of Japanese Affiliated Automobile Parts Manufacturers*, Tokyo: Sakura Research Institute.

Nexus Associates, Inc. (2000) *Evaluation of Industry Technical Institutes in Thailand*, report submitted to the Royal Thai Government and the World Bank, October.

Poapongsakorn, N. (1997) 'ASEAN Automobile Industry in the Emerging Economic Integration Environment: Corporate Strategies and Government Policies'. A paper presented for the 2nd Export Meeting of the Interregional Project on 'Transnational Corporations and Industrial Restructuring in Developing Countries' organized by Division of Investment, Technology and Enterprise Development, UNCTAD, Geneva, November.

Poapongsakorn, N., Chernsiri, C., Naiwithit, V., Wattanachit, S., Veerakul, K. and Panyaswadsut, C. (1992) *Training in the Manufacturing and Service Sectors*, research report prepared for the National Economics and Social Development Board, Faculty of Economics, Thammasat University, February (in Thai).

Poapongsakorn, N. and Wangdee (2000) 'The Thai Automotive Industry'. The ASEAN Automotive Industry: Challenges and Opportunities, A paper submitted to Australian Pacific Economic Cooperation Committee, Australia.

Ritchie, B.K. (2001) 'The Political Economy of Technical Intellectual Capital Formation in Southeast Asia', unpublished Ph.D. dissertation, Department of Political Science, Emory University.

Romijn, H. *et al.* (2000) 'TNCs, Industrial Restructuring and Competitiveness in the Automotive Industry in NAFTA, MERCOSUR and ASEAN', United Nations Conference on Trade and Development, *The Competitiveness Challenge: Transnational Corporations and Industrial Restructuring in Developing Countries*, New York and Geneva: United Nations.

Thailand Development Research Institute (TDRI) (1998) 'Manpower Development Approach for Long-Term Industrial Development', final report prepared for the Office of Industrial Economics, Ministry of Industry, Royal Thai Government, December.

United Nations Conference on Trade and Development (UNCTAD) (2000) *The Competitiveness Challenge: Transnational Corporations and Industrial Restructuring in Developing Countries*, Geneva: United Nations.

Appendix 9.A1 OJT and Off-JT: bivariate probit regressions for all industries

Variable	All workers				Non-production			
Independent var.:	Dep. var.: OJT		Dep. var.: Off-JT		Dep. var.: OJT		Dep. var.: Off-JT	
	Robust coeff.	z	Robust coeff.	z	Robust coeff.	z	Robust coeff.	z
Years of schooling	−0.677143	−0.935	−0.0167047	−0.265	−0.4544513	−1.664	−0.2504331	−0.909
Year of schooling squared	0.0006762	0.212	0.0011733	0.398	0.0164037	1.483	0.011471	1.042
Age	−0.035429	−1.197	0.0098488	0.432	−0.0232271	−0.244	0.0269713	0.322
Age squared	−0.000305	0.633	−0.0000734	−0.178	0.0002355	0.161	0.0003164	0.247
Tenure	−0.0307449	−1.745	0.0284051	1.721	−0.0797418	−1.872	0.033819	0.832
Tenure squared	0.0007897	0.958	−0.0006034	−0.61	0.0030067	1.918	−0.0014497	−0.648
Male	−0.169109	−1.815	0.0048148	0.05	−0.2638615	−1.638	−0.5693991	−1.561
Position_dummies = 6					position_dummies = 3		position_dummies = 3	
Time_dummies = 2					time_dummies = 2		time_dummies = 2	
Employment_dummies = 19					employment_dummies = 19		employment_dummies = 19	
Constant	1.9539630	3.276	−0.1120005	−0.253	4.3666000	3.074	1.570057	0.642
Athrho	0.2426412	9.862			0.4148759	6.835		
Rho	0.2379889				0.0392605			
Number of obs. = 4,691					Number of obs. = 1,000			
Wald chi² (16) = 732.89					Wald chi² (17) = 5,692.70			
Log likelihood = −5177.8642					Log likelihood = −958.89966			
Wald test of rho = 0					Wald test of rho = 0			
chi² (1) = 97.2494					chi² (1) = 15.6683			

Source: Calculated from TDRI, *Education and Training Survey* (August 2001)

Appendix 9.A2 OJT and Off-JT: bivariate probit regressions for the auto parts industry

Variables	All workers				Non-production			
Independent var.:	Dep. var.: OJT		Dep. var.: Off-JT		Dep. var.: OJT		Dep. var.: Off-JT	
	Robust coeff.	z	Robust coeff.	z	Robust coeff.	z	Robust coeff.	z
Years of schooling	-0.3755096	-2.315	-0.1881859	-1.588	-1.9470050	-2.778	0.3567483	0.540
Years of schooling squared	0.0129924	1.870	0.0077583	1.283	0.0760285	2.798	-0.0156747	-0.608
Age	0.1155823	1.565	0.1760427	4.407	0.1150050	0.006	-0.0934666	-0.552
Age squared	-0.0025531	-2.050	-0.0030243	-6.336	-0.0019422	-0.542	0.0018724	0.865
Tenure	-0.3777940	-1.541	0.0675453	4.197	-0.0660665	-0.772	0.1138673	2.237
Tenure squared	0.0023871	2.618	0.0003833	0.287	0.0040191	2.306	-0.0042370	-1.863
Male	-0.0213745	-0.281	0.3653602	5.309	0.1104992	0.345	-5.8514490	-1.021
Position_dummies = 6					position_dummies = 3			
Time_dummies = 2					time_dummies = 2			
Employment_dummies = 5					employment_dummies = 5			
Constant	1.2961660	0.804	-1.7670710	-2.837	11.9119200	5.078	5.9734090	—
Athrho	0.3109022	8.388			0.5103433	—		
Rho	0.3012576				0.4702126			
Number of obs. = 1,144					Number of Obs. = 237			
Wald chi² (4) = 0.49					Wald chi² (4) = 19.81			
Log likelihood = -1263.9732					Log likelihood = -257.09567			
Wald test of rho = 0					Wald test of rho = 0			
Chi² (1) = 70.3601					chi² (1) = 71.5404			

Source: Calculated from TDRI, Education and Training Survey (August 2001)

10 Producing auto parts in Malaysia

Skill formation in forging and casting

Yuri Sadoi

The Malaysian government started to promote the automobile industry in the 1980s in order to develop the technical capability of the Malaysian production industry. Malaysian industrialization went through four distinct stages between 1967 and the 1990s (Alavi 1996): (1) import substitution industrialization between 1957 and 1970; (2) export-oriented industrialization between 1970 and 1980; (3) import substitution industrialization, especially to promote heavy industry, between 1980 and 1985; and (4) the current push for further export-oriented industrialization. During the third stage, the Malaysian government promoted the automobile industry, hoping that it would foster higher economic growth through stronger backward and forward industrial linkages. Within the Malaysian automobile industry, knowledge in basic metals and engineering were thought to be particularly essential for technological development. The automobile industry involves material producers, partmakers and assemblers spread out over various industries.

In 1983, the first national car company, Perusahaan Otomobil Nasional Sdn. Bhd. (Proton), was established as a joint venture between Heavy Industry Corporation of Malaysia (HICOM) and Mitsubishi, and in 1985 Proton started automobile production. By enjoying tax benefits and tariff protection, the auto industry in Malaysia both developed and increased its production over the following fifteen years. In 1994, the second national car company, Perodua, was established with the Japanese maker Daihatsu. By 1997, the two national car companies, Proton and Perodua, increased their production volume to 280,000 units and together shared over 80 per cent of total sales in Malaysia.[1] Although the Malaysian automobile industry suffered during the economic crisis that started at the end of 1997 and lasted throughout 1998, it showed a quick recovery from 1999 onwards. According to the Malaysia Automobile Association, automobile sales in Malaysia during 2000 recovered to 85 per cent of the sales peak from 1997.

Over a period of nearly two decades, the Malaysian auto industry had developed rapidly, increasing its production volume and improving the rate of local parts content. Malaysia, however, has yet to meet the requirement of the international organization to which it belongs to open its market for trade and investment. The Association of South East Asian Nations (ASEAN) has agreed to slash tariffs on foreign cars to between 0 and 5 per cent by

2003. Since the economic crisis in 1997, strong doubts remain as to whether all member states will be willing and able to refrain from protectionism. Malaysia requested a two-year reprieve before being required to cut tariffs aimed at protecting its local car industry. The establishment of free trade areas and deregulation will develop into a global sourcing of auto parts. Local content regulation must be abolished, and indeed Malaysia no longer offers incentives based on local content. As a result, local suppliers have to improve quality, lower costs and shorten delivery time in order to survive the growing competition from foreign suppliers.

Technology transfer from the Japanese auto industry has played an important role in the development of the Malaysian auto industry. Production technology and operational expertise have been transferred to the two Malaysian national car companies, Proton and Perodua, from their joint venture partners, Mitsubishi and Daihatsu, respectively. At the same time, under government sponsorship, Proton started a local vendor development programme that matched local vendors, preferably Bumiputeras,[2] with overseas vendors, many of whom were Japanese suppliers. To establish a local capacity to produce auto parts, various kinds of production technology have been transferred to local suppliers.

In order to transfer the Japanese production system, the Japanese skill training system was implemented in Malaysia. The accumulation of technical knowledge and skills on the part of engineers, technicians and workers is the key factor for upgrading a country's technological capability (Koike and Inoki 1990). However, the Japanese skill formation system, developed under the long-term investment in employees as secured by the lifetime employment system, was not as effective in Malaysia as in Japan. Both Malaysian industry and the national government criticized and complained about the slow transfer of Japanese technology to local operations (Jomo 1994).

The focus of this chapter is on forging and casting, skills that are used in producing major engine and transmission parts. Forging is a method used to shape induction-heated steel or aluminium by machine hammering, pressing and rolling. Casting is a method by which to shape molten iron or aluminium by pouring it into a metal mould. Few studies have been conducted on the importance of the skills involved in forging and casting with regard to the upgrading of technology in host countries. Forging and casting skills are, in fact, difficult to transfer because, in most cases, suppliers in Japan are small- or medium-sized firms in which on-the-job training (OJT) is the main source of skill acquisition. The first part of the discussion that follows involves the importance of mastering the skills for forging and casting within the context of local capacity for production. Then, processes involved in acquiring skills in forging and casting are introduced. With reference to some successful cases of technology transfer, I will present an analysis of how high-level expertise and skills were transferred to workers and technicians in Malaysia. Although the number of skilled workers and technicians in forging and casting are limited, the cases discussed here will show how workers in Malaysia can be upgraded and how the Malaysian automobile industry can be developed into an internationally competitive one.

The status of forging and casting parts locally

In order to explain technical capability in the auto parts industry in Malaysia, the progress made in manufacturing parts locally for Proton cars is used as an indicator of the local suppliers' level of technological capability. Since Proton started production in 1985, production volume has increased and the ability to produce parts locally has improved dramatically. In 1996, the local content rate had risen, reaching nearly 80 per cent based on local material content points.[3] The local material content policy lists 314 parts and assigns them scores that add up to 115 points in total.

An automobile consists of thousands of kinds of parts. Each automobile part has its own production process, technique and requirements. In order to explore the progress of technology transfer and the current level of technical ability in automobile parts production in Malaysia, it is necessary to analyse the manufacturing techniques for major automobile parts.

Table 10.1 summarizes the production techniques of major parts and the status of local production capability in Malaysia during 2000. From the perspective of production techniques involved, parts production can be divided into three categories: (1) assembling from sub-parts; (2) casting from imported moulds; and (3) utilizing multiple production processes. With regard to the first, most assembly-oriented production parts were already made locally. Before the National Car Project (NCP) started, the majority of parts were produced by assemblers importing most of the necessary components rather than by manufacturers. The second technique, for making body parts, utilizes metal moulds by stamping or pressing, or plastic injection moulding. Most of these processes are done at the local level, but the metal moulds are generally imported from technical assistant partners. The third category includes multiple production processes of casting and machining, die casting and machining, and forging, heat treatment and machining, which are all very difficult to realize on the local level. These production processes are used in the production of engine and transmission parts.

While most of the body parts were already locally produced, the local manufacture of engine and transmission parts still lags behind. Sadoi (1998) pointed out three reasons for this: (1) lack of scale economies; (2) low designing capability; and (3) technical difficulty. Thus the first problem encountered when parts are locally produced is a low production volume, and a low production volume does not meet the requirement of scale economies. A detailed description of this problem will be skipped here because it is not related to human skills.

The second involves functional parts,[4] the parts that perform high technical functions. As the automobile has increased in complexity, these functional parts have come to play a more important role in automobile technology. Automobile makers order various parts by simply specifying their functions and their positions within the entire layout, information that the parts manufacturers use to develop designs and produce them. Producing these functional parts on a local level is difficult in that their production requires the supplier to possess engineering skills that enable both the interpretation of the

Table 10.1 Major manufacturing processes and status of local production capability in Malaysia, 2000

	Production techniques	Major parts	Examples	Status of local production capacity
1	Assembling	Electrical parts	Radio, lamp, battery, wire harness	100–75 per cent
2	Pressing	Body parts	Body frame, floor, door, bonnet,	100–75 per cent
	Plastic injection moulding		Bumper, dashboard	100–75 per cent
	Glass tempering		Glass	100–75 per cent
3	Casting and machining	Engine,	Cylinder block piston, flywheel,	75–50 per cent
	Die-casting and machining	transmission and	Water pump, fuel pump	50–25 per cent
	Forging and machining	chassis parts	Drive shaft	25–20 per cent

Source: Personal communication, Mitsubishi Motors Corporation to Y. Sadoi (1998, 1999, 2000)

functional specifications detailed in the auto makers' orders and the subsequent design process. This is a kind of joint product development between the automobile maker and the parts supplier in the sense that both are involved in the development of new products. This sort of interaction does not appear to take place in Malaysia. Functional parts are still imported, or only the final assembly process is carried out locally, as most of the partmakers in Malaysia are not yet capable of designing.

Technical difficulty is cited as a third obstacle for local manufacturers in undertaking certain production processes and meeting accuracy requirements. Most of the imported parts necessitate more than three major production processes and require high accuracy and durability because they are used in the making of engines or transmissions. In addition, most foreign-made parts involve forging, casting and multiple machining processes in their manufacture.

Table 10.2 shows major engine parts, as well as their major production processes and local manufacture status. As shown here, most of the

Table 10.2 Production processes and status of local production capability of major engine parts in Malaysia, 2000

Part	Cast	Die.C	Forg	Press	Heat	Mach	Plas	Assy	Funct	Local
Intake/exhaust valve			*		*	*				
Con-rod			*		*	*				
Crankshaft			*		*	*				
Cam shaft	*					*				
Carburettor/ fuel injector		*				*		*	*	
Fuel pump		*				*		*	*	
Catalytic converter			*					*	*	
Turbo charger	*					*		*	*	
ECU (computer)									*	
Sensors									*	
Water pump		*				*		*	*	
Cylinder head	*	*				*				L
Cylinder block	*					*				L
Piston	*				*	*				L
Timing belt cover							*			L
Timing belt							*			L
Flywheel	*					*				L
Oil pump		*								L
Oil seal		*								L
Fuel tank				*						L
Intake manifold	*					*				L
Exhaust manifold	*					*				L
Fan							*			L
Radiator			*			*		*		L

Source: Personal communication, Mitsubishi Motors Corporation to Y. Sadoi (2000)

Note

Die.C: diecast; Mach: machining; Plas: plastic injection; Assy: assembling; Funct: functional parts; L: locally produced parts

foreign-made engine parts, such as valves, con-rods and crankshafts, or parts with highly technical functions, such as fuel pumps and sensors, need forging, heat treatment and machining processes. As far as workers' skills are concerned, those needed for producing parts with high accuracy and durability fall under the heading of activities such as forging, casting and precision machining.

The availability of forged and cast parts is low not only in Malaysia, but also in other ASEAN countries. In 1996, 15 and 28 per cent of cold forging and hot forging parts were purchased in ASEAN countries, while nearly 50 per cent of steel and aluminium cast parts were available in ASEAN countries (Sakura Sogo Kenkyusho 1996: 23). Hot forging is a method used to shape induction-heated steel by machine hammering, pressing and rolling. Cold forging is essentially the same method, but instead the steel is shaped without induction heating. Forged steel has high tensile strength and can withstand repetitive impact in an engine; therefore, it is used for the engine's main moving parts. Hot forging is used to make the parts that undergo bigger transformations but require less accuracy, while cold forging is used to make the parts that undergo smaller transformation but require higher accuracy. Because of the accuracy requirements, cold forging is the more difficult technique. Aluminium too is shaped by hot and cold forging, but it is a more difficult material to work with than steel because temperature and pressure require stricter control.

Skills training

The local manufacturing of parts has been promoted since the beginning of the National Car Project. To meet the demands for locally produced parts, many foreign suppliers have preferred to set up joint ventures with Bumiputeras. Local auto parts suppliers received technical assistance from their foreign counterparts and, as a result, Malaysian workers were urgently required to develop high skills to produce the parts.

To increase workers' skills, the Malaysian government set up several training institutes. Skills training in the public sector was thenceforth conducted by the following four agencies:

1   The Ministry of Education is in charge of secondary vocational and technical education. Enrolment in upper secondary education was at a level of about 50 per cent in 1995. From a total of 514,970 students enrolled in upper secondary education, 3,386 students were in technical and 29,083 were in vocational schools.[5]

2   The Ministry of Human Resources gives skill certifications through the National Vocational Training Council and operates training institutions such as the industrial training institutions (ITIs) and the Center for Instructor and Advanced Skill Training (CIAST), which provide training in high technical skills for instructors, and diploma courses. Sponsored by the Japanese government under the ASEAN Human Resources Development Project, CIAST was established in 1984.

3   The Ministry of Youth and Sports is in charge of youth training centres that provide advanced training centres and vocational training to young people between the ages of 16 and 26.
4   Majlis Amanah Rakyat (MARA) is in charge of the Pusat Giat skill development centres (GIAT) and the MARA Vocational Institute (IKM). GIAT and IKM provide training in the practical and theoretical skills needed in industry.

Although the Malaysian government is setting up various skill-upgrading programmes, it still faces a shortage of skilled workers, technicians and engineers. According to the Center of Educational Excellence Malaysia in 1997, the demand for skilled workers was about 270,000, while there were only about 110,000 available at that time. About 48,800 engineering assistants were needed, while only roughly 22,900 were in supply. As for engineers, there were about 5,000 available, with about 36,300 in demand.

The supply and demand imbalance is not the only problem. Certain kinds and levels of skills needed presented a serious bottleneck. The skill certification system (SKM) was initiated in 1973 in order to promote vocational training. Five levels were established, the lowest being level 1 and the highest, level 5. Levels 1 and 2 concern operation and production, level 3 is supervisory, level 4 is equivalent to a diploma in technology for supervisors and managers and finally, level 5 is equivalent to a degree in technology for managers.

The number of certificate holders increased: in 1995, over 25,000 workers received SKM skill certificates, and by 2001 this number had risen to 44,288.[6] Among these recipients, however, most have operation and production levels, some attained supervisory levels, and only a few achieved management level. In 2001, SKM level 1 certification was issued to 16,577 workers, level 2 to 21,157 workers and level 3 to 6,554 workers. There were no recipients at levels 4 or 5. The level 3 certificate recipients between 1992 and 1997 in tool and die making numbered forty-five; nineteen in general machining; fifty-one in machine fitting; twenty in stamping; and 140 in welding.[7] There were no certificate recipients in forging and accuracy machining, areas where skilled workers are urgently required.

In order to promote enterprise-based training, the Malaysian government implemented two incentives for human resource development (HRD). In 1987, the Double Deduction Incentive (DDIT) and, in 1993, the Human Resource Development Fund (HRDF)[8] were introduced. In 1993, only 8.3 per cent of firms used the DDIT (World Bank 1997). In 1994, a total of 3,417 employers were registered with the HRDF and a levy of MYR 88 million was collected, but only 30 per cent of the total levy collected was used for training grants.[9] In total, between 1993 and 2001, MYR 1 billion in grants was allocated to the Human Resources Development Council for the purpose of training and upgrading skills.

Enterprise-based training

In Malaysia, there are several obstacles to skills upgrading for workers, such as insufficient government support, employers' lukewarm attitudes, an unfavourable skills environment and weak interest in upgrading among individuals (Sadoi 1998). I wish to concentrate here on enterprise-based training and examine the skill-formation processes in firms.

Enterprise-based training takes three main forms: (1) off-the-job training (Off-JT), which enables workers to take short training courses either in-house or at external training centres; (2) OJT, which lets workers receive training at production sites under the supervision of skilled workers and technical advisers from overseas; and (3) overseas training, which provides the operational and maintenance training that cannot be done effectively at home because of a lack of necessary machinery or technical advisers.

According to the Malaysia Industrial Training and Productivity (MITP) survey, which covered 2,200 manufacturing firms in 1994 and 1995, about 20 per cent of firms had enterprise-based Off-JT and most of them (17 per cent) combined it with OJT. About 45 per cent of firms gave only OJT to their workers, while 35 per cent gave no training at all. The proportion of firms giving Off-JT increases with firm size (World Bank 1997).

Of the fifteen auto parts suppliers interviewed in the years between 1997 and 2000,[10] all were found to give their workers both Off-JT and OJT, and all were found to give new workers introductory Off-JT that explains working conditions, service, safety, health care and basic technical terms in use at the firms. After this, basic operational know-how was taught through OJT only. One supplier provided an annual training programme for workers, but all the others gave Off-JT on an ad hoc basis. Among the total of 212 workers surveyed in those fifteen suppliers,[11] 35 per cent of them indicated that they had received at least one session of Off-JT training. Among those who received Off-JT, fourteen took in-house quality control and ISO 9000 courses, thirteen took technical training at external training centres, twelve took in-house computer training, seven took in-house supervisory training, three took their training overseas and three took in-house courses in machine operation. The rest took in-house courses in fire drills, safety and language. As shown here, only some selected workers had opportunities for Off-JT. Opportunities for overseas training were prepared by six out of fifteen suppliers, but, from among 212 workers surveyed, only three had had overseas training. All the suppliers had, for certain periods, technical advisers from Japan or other foreign countries. As shown above, although all the suppliers prepared training programmes for workers, the likelihood of any particular worker receiving Off-JT were low and OJT was provided when necessary on an ad hoc basis.

Casting

Skill formation processes in casting at plant level are introduced in this section, in which two cases are presented where the casting of parts locally and the skill acquisition process in casting plants are examined. The casting process is used for making major engine parts. In order to increase the rate at which engine parts are produced locally, HICOM Engineering Sdn. Bhd. was incorporated in 1984 and started production in 1991. The following year, Proton started its casting project, building a state-of-the-art casting plant and training workers for it.

HICOM Engineering Sdn. Bhd. was incorporated in 1984 by HICOM, whose central aim was to pilot Malaysia's industrialization programme. It started its foundry operation in 1991 and its machining operation in 1993. It produces exhaust manifolds, flywheels and engine support brackets for Proton cars, as well as motorcycle crankcases and brake drums for other customers.

The organization of the shop floor is shown in Table 10.3. The qualifications for foremen, assistant foremen and line leaders are SPM, SPVM, IKM, or ITI certifications (see list of abbreviations at the start of the book). At least five years of experience in a related area is required for foremen, and three years for assistant foremen and line leaders. It is preferred that operators have an SPM, SPVM, IKM, or ITI certification, but no further experience is needed. Operators are graded into four levels (grades IV–I). The promotion criteria are qualifications, merit, seniority, ability, technical capability, attitude and loyalty to the company. Outstanding operators can be promoted in the following manner: from grade IV to grade III after a minimum of two years of service; from grade III to grade II after two and a half years; and from grade II to I after three years. It is worth noting that the majority of operators in grade IV are foreign workers.

All new employees receive introductory training that covers the terms and conditions of service, ISO awareness and safety and health care. After this, workers receive Off-JT and OJT. Some selected workers and foremen receive Off-JT in a leadership training course and a technical upgrading course either in-house or at CIAST. Others acquire operational know-how during their OJT. There was, for example, a technical adviser dispatched from a casting parts supplier in Japan who was stationed in the Malaysian plant for two years to teach and advise foremen and workers during OJT when advice was necessary. It is further policy that whenever new technologies and machinery are introduced, workers in charge of the relevant processes take a one-month training period in Japan at the casting parts supplier with which the plant has technical agreements. For instance, one Malay foundry manager who had completed a one-year training in intensive casting at the Mitsubishi Motors casting plant in Kyoto now plays an important role in providing maintenance and quality control expertise for the OJT of other workers.

Casting at Proton

A casting plant for major Proton engine parts was launched in 1992 and started its operations in 1994. It was decided that two major engine parts (the cylinder block and bearing cap), brake disks and brake drums were to be

Table 10.3 Organization in casting

	Foreman	Asst foreman	Line leader	Operator			
				I	II	III	IV
Qualification	SPM, SPVM, IKM, or ITI			SPM, SPVM, IKM, or ITI is preferred			
Experience (years)	5	3	3	No requirement			
Minimum years for promotion		3		3	2.5	2	
Number of workers (foreign workers)	5	3	22	90			2 (121)

locally produced in this new casting plant by the end of 1994. During the preparation period, managers, engineers, supervisors, technicians and workers received training in a casting plant in Japan. None of them had any previous working experience in or knowledge about casting. Fifteen managers and maintenance workers were sent to Japan, where they received eighteen months of training at a casting plant. They received instruction in theory, were exposed to practical training and learned about personnel management. After a brief orientation course, they completed one and a half months of lectures on casting techniques and equipment, and one week each on purchasing, prototype making, casting and maintenance. After these courses, they received OJT in technical procedures, such as melting, sand preparation, core making, moulding and finishing.

From 1993 to 1994, twenty-three foremen and line leaders received ten months of training in Japan, and twelve workers received eight months of training. After a month of introductory theoretical training, trainees received technical instruction in their specific areas: two foremen in melting, two in moulding, two in sand preparation, two in core making, one in finishing and six in maintenance. They trained by OJT using similar machinery and under similar production conditions to those already installed in the new casting plant in Malaysia. All the initial foremen and core workers in each process received six months to one year of intensive training overseas before production started. In total, sixty people underwent training overseas during the preparation periods between 1992 and 1994 (Sadoi 2003).

How do high-level workers obtain their skills in Malaysia? In order to examine skill formation on an individual level, we will take a closer look at the skill acquisition process of one particular assistant foreman at casting. This assistant foreman worked at the melting group of the Proton casting plant and obtained his skills through the intensive training programme in Japan. A graduate of upper secondary school and with an SPM, his first working experience was as a mechanic at several car repair shops. When Proton started its assembly plant in 1985, he began as an operator, at the lowest grade, of a stamping machine for small body parts. He continued in this job for eight years. In 1993, the entire stamping shop was transferred to a new local vendor. He had the choice either to move to this new local vendor or to transfer to another section within Proton. At that point, he was asked to transfer to the new casting plant for the melting shop and to take a one-year training programme in melting in Japan. He decided to accept the last option, as he knew that training in Japan would upgrade his status in his new section. In Japan, he took intensive lessons in Japanese, a lecture in basic casting and practical instruction in melting. The daily training schedule consisted of a one-hour lecture plus five hours of OJT concerned with melting, furnaces and maintenance at production sites, and a two-hour Japanese lesson. There was an individualized approach to the practical training because he was the only trainee at the foreman level in the discipline of melting. A Japanese engineer in casting was in charge of theoretical lectures, while an assistant foreman from the melting shop gave him OJT on furnaces, melting and maintenance. After completion of his training in Japan, he was promoted to assistant foreman in melting. Together with some Japanese technicians during the seven months of preparation before the casting plant started production, he assisted with the pre-production testing as a leader of melting. He was involved in machinery installation and the training of new workers in the melting group. Among the ten workers in his group, only one had received training overseas in melting, while the others, transferred from other Proton departments or newly hired, had no experience in or knowledge about casting. The casting plant started production after he had completed a total of eighteen months of training (comprising eleven months in Japan and seven months of preparation in Malaysia). Initially, a Japanese adviser helped him in most operations. His work routine began with a morning meeting with workers, after which he checked the conditions of workers, lines and machinery to keep production smooth and safe. During the initial stage, he relied on the Japanese adviser when troubles occurred. Troubleshooting formed a good OJT experience for both him and his co-workers (Sadoi 2003).

Forging

The skill-formation processes in forging at plant level in Malaysian firms are examined in this section within the context of the process of creating a local capability for forging parts. As described in an earlier section, the rates at which local capability for production are achieved in the areas of casting and forging are still low in Malaysia. Forging is the least available skill not only in Malaysia, but also in other South-East Asian countries. There are some cases, however, in which the forging process has been transferred successfully to Malaysia.

Clutch discs, which require forging to produce, began to be manufactured locally in 1997. The clutch discs are the basic type of disc commonly used for several types of manual transmission. In 1995, Exedy (Malaysia) Sdn. Bhd. (Exedy M) was incorporated in Nilai, Negeri Sembilan with the equity of Exedy Corporation Japan (50 per cent), Proton (45 per cent) and Yew Teong (M) Sdn. Bhd. (5 per cent). Exedy M installed production lines in 1996 and started production of the clutch disc. By 1997, local production had reached 34 per cent. Some of the major components of the disc required forging, heat treatment and precision machining, and Exedy (Japan) did not have those skills in-house, rather opting to subcontract those processes. In Malaysia, however, no such proper subcontractor was available. To create local capabilities for these processes, subcontractors in Japan who built forges, heat treatment services and precision machines were asked to establish production plants in Malaysia. YMC Forging for hot and cold forging of parts, Tohken (M) for heat treatment services and Rivatec (M) for machining, pressing and broaching processes were established in 1996 and started operations in 1997. It was important that all three makers were set up at the same time. Had even one maker declined, the whole project would have failed.[12] Each supplier planned its own workers' skill training programmes.

YMC Forging was set up with management gleaned from its majority shareholder, a forging company in Japan: a Japanese chairman, a Japanese managing director to oversee the administration department and a Japanese director to manage the production, sales engineer and die and mould departments. They were joined by one Malay director, one Malay assistant manager, six Malay staff members and thirteen workers (of whom nine were Indians and four Malays). Operational and skills training were planned and implemented by the managing director and the director. Their training aims were, first, to develop a Malay assistant manager, and then, to have the assistant manager train workers.

Even though there have been few specialists in the area of forging in Malaysia, the following case of a Malay worker shows how a successful skill acquisition process resulted in his achieving assistant manager status in a forge maker. This particular worker, having attained the highest grade level by the end of his secondary education, received a four-year scholarship in Japan. During his first year in Japan, he took a one-year intensive Japanese language programme in Osaka. In the following year, he was admitted to Yatsushiro Technical College in Kumamoto Prefecture in Japan, which has a five-year programme: three years at high school level, followed by two years at junior college level. He was admitted to the mechanical–electrical engineering programme at the third-year high school level. In his final year at the college, he took a required two-week company training course. He chose to train for the forge maker because he heard that the maker would start YMC Forging in his home town in Malaysia. In addition to the two-week training as part of the college programme, he was approved for a five-month practical training period under the auspices of the Association of Overseas Technical Scholarship because the forge maker had been looking for a key technician for its new plant in Malaysia.

The training he received in forge making was practical, intensive and specially designed so that he could become a key technician in the specific plant, but in no way different from how Japanese trainees were given instruction. The forge maker, small in terms of its fifty-eight employees, and having evolved from a small-scale forge maker into a modern factory, did not have a formal training programme. As in the case of other small manufacturers in Japan, skills were transferred mainly by OJT. He gained operational and maintenance expertise in machinery to be installed in the new plant in Malaysia, such as the air stamp hammer, air hammer, knuckle press, crank press, billet shearing machine, induction furnace, shop blasting machine and measuring apparatus. He worked on one machine for a few weeks under a Japanese supervisor, and then switched to another. After he had acquired the necessary operational and maintenance expertise, he attended a brief lecture about metal mould making.

On the completion of his four-year scholarship in Japan, he was given employment as a technician at YMC Forging in Malaysia at the time the new plant was being laid out. His job involved setting up machinery and preparing the operation together with Japanese technicians. Through the entire process of starting up a plant and OJT, he learned about production engineering, productivity improvement, quality control, maintenance, as well as about how runs are tested and operators are trained. His involvement in the start-up process made him a key Malaysian technician in the new plant. After production started, he became an assistant manager in charge of quality control inspection, the engineering service line and the die and mould departments (Sadoi 2003).

In short, he mastered his forging knowledge only through intensive OJT. The forging technique is in fact difficult to describe in a manual, such as the manner by which temperature can be ascertained by watching the colour change on the surface of forged iron. OJT is thus the major method of skill acquisition in forging.

Conclusion

This chapter discussed the skill-formation processes in Malaysia in the area of auto parts production. To judge from the progress made locally in various auto parts production technologies, it appears that local manufacturing capabilities in that area are lagging behind. In particular, forging and casting are indispensable processes involved in the production of engine and transmission parts.

Realizing the importance of HRD, the Malaysian government has set up training institutions and incentives for employers to train their workers. Malaysia has to develop industries that produce higher value-added products, but it faces a shortage of skilled workers and technicians with which to do so.

Forging technology was transferred to Malaysia in recent years. Only a few small, modern forge makers have started production, and few forging courses are available in training centres. Therefore, OJT – both in-house and at the mother company in Japan – is the only method by which to acquire forging skills. The combination of intensive overseas training and practical OJT seems effective for building up operational and maintenance expertise in forging. In particular, OJT during the start-up phase of a new plant, which includes setting up machinery and pre-production testing, was shown to be essential in terms of the development of key skills and knowledge. The forging and casting processes are important in the upgrading of Malaysia's production capability aimed at rendering its automobile industry internationally competitive.

Notes

1 Proton company reports.
2 Indigenous Malaysian.
3 Proton corporate profile. The percentage is based on the highest point model.
4 Clark and Fujimoto (1991) defined these kinds of parts as black box parts.
5 Malaysia (1996) and Malaysia, Department of Statistics (1997).
6 Malaysia, National Vocational Training Council, Ministry of Human Resources (2002).
7 Malaysia, National Vocational Training Council, Ministry of Human Resources (1998).

8 An employer with fifty or more employees in a manufacturing industry has to register with the HRD council and pay an HRD levy at the rate of 1 per cent of monthly salaries to HRDF. When the employer enrols participants in HRDC approved courses, the employer enjoys large discounts or a rebate on fees.
9 *Human Resources Development and Training Guide 1995.*
10 Interviews conducted by the author between 1997 and 2000 at auto parts suppliers in the Selangor area.
11 Surveyed by the author between 1999 and 2000 at the fifteen auto parts suppliers in the Selangor area.
12 Information obtained from managers of Exedy (M), YMC Forging, Tohken (M) and Rivatec (M); interviewed by the author 1998–1999.

References and further reading

Alavi, R. (1996) *Industrialization in Malaysia: Import Substitution and Infant Industry Performance*, New York: Routledge.

Challenger Concept (M) Sdn. Bhd. (1995) *Human Resource Development and Training Guide*, Kuala Lumpur.

Clark, K.B. and Fujimoto, T. (1991) *Product Development Performance*, Cambridge, MA: Harvard University Press.

Fujimoto, T. (1995) 'A Note on the Origin of the "Black Box Parts" Practice in the Japanese Motor Vehicle Industry', in Shiomiannd, H. and Wada, K. (eds) *Fordism Transformed*, New York: Oxford University Press.

Jayasankaran, J. (1993) 'Made in Malaysia: The Proton Project', in Jomo, K.S. (ed.) *Industrializing Malaysia*, London: Routledge.

Jomo, K.S. (ed.) (1994) *Japan and Malaysian Development: In the Shadow of the Rising Sun*, London: Routledge.

Koike, K. and Inoki, T. (eds) (1990) *Skill Formation in Japan and Southeast Asia*, Tokyo: University of Tokyo Press.

Malaysia (1996) *Seventh Malaysia Plan 1996–2000*, Kuala Lumpur: Government Printer.

Malaysia, Department of Statistics (1993) *Yearbook of Statistics Malaysia 1992*, Kuala Lumpur: Government Printer.

—— (1995) *Yearbook of Statistics Malaysia 1994*, Kuala Lumpur: Government Printer.

—— (1997) *Yearbook of Statistics Malaysia 1996*, Kuala Lumpur: Government Printer.

Malaysia, National Vocational Training Council, Ministry of Human Resources (1994) *List of National Occupational Skill Standard* (mimeographed), Kuala Lumpur: Government Printer.

—— (1998) *Report of Skill Certification Record* (mimeographed), Kuala Lumpur: Government Printer.

—— (2002) *Report of Skill Certification Record*, Kuala Lumpur. Online. Available at http://www.nvtc.gov.my (accessed 2 October 2002).

Pillai, P. (ed.) (1994) *Industrial Training in Malaysia: Challenge and Response*, Kuala Lumpur: ISIS Malaysia.

Sadoi, Y. (1997) 'Auto Parts Localization in Malaysia', *Journal of International Business and Entrepreneurship* 5 (2): 51–98.

—— (1998) 'Skill Formation in Malaysia: The Case of Auto Parts Industry', *Southeast Asian Studies*, 36 (3): 317–354.

—— (2003) *Skill Formation in Malaysian Auto Parts Industry*, Bangi: Universiti Kebangsaan Malaysia Press.

Sakura Sogo Kenkyusho (1996) *Development and Intra-ASEAN Cooperation of the Auto Industry in Southeast Asia*.

World Bank (1997) *Malaysia: Enterprise Training, Technology, and Productivity*, Washington, DC: World Bank.

Unpublished internal company documents

Proton (1997) *Proton Corporate Profile.*
—— (1998) *Proton Corporate Profile.*
—— (1999) *Proton Corporate Profile.*
—— (2000) *Proton Corporate Profile.*

11 The Toyota Production System in Indonesia

Keisuke Nakamura

This chapter examines how workers' groups contribute to the smooth running of the Toyota Production System (TPS) at Toyota-Astra Motor (TAM) operating in Indonesia, and also how human resource management (HRM) practices succeed in obtaining such workers.[1]

TAM, which is jointly owned by Astra International (51 per cent) and Toyota Motor Corporation (TMC) (49 per cent), has been operating in Indonesia for about thirty years and has grown from being just an importer and distributor to an automaker with functions such as welding, stamping, casting and assembling. TAM has been transplanting the TPS gradually and successfully. While six Japanese-related automakers together have an approximate 90 per cent share in the Indonesian automobile market, TAM has held the first or second position with a relatively stable market share of around 25 per cent.[2]

The smooth running of the TPS depends partially, but deeply, on workers' groups with intellectual skill (Koike 1999: 11–16). The TPS is characterized by adherence to observing a production plan, orientation towards continuous improvement and flexibility from the perspective of production management. While many factors usually disturb daily production in terms of schedule, cost and operation rate, engineers, line managers, foremen and workers under the TPS are expected to keep the plan strictly on track. Once they discover any trouble on a production line, they trace the causes and take the necessary measures as soon as possible. They also try to find better ways of line layouts and work methods, as well as better tools and jigs. They are expected to be continuously improving the line so as to decrease defect rates and reduce costs. The TPS is flexible in that takt time (production line speed or conveyor speed) can be adjusted to changes in production volume, and several types of vehicles can be produced on the same production line without influencing the speed of the line. At Toyota, this is called flexible levelled production. Engineers, line managers, foremen and workers are together responsible for running the flexible TPS. While engineers and line managers play a major role in managing the TPS successfully, workers' groups with intellectual skill led by foremen are also expected to help facilitate the smooth running of the TPS on the shop floor.

HRM practices should aim towards recruiting these workers and eliciting more effort from them. Workers need on-the-job and off-the-job (OJT and

Off-JT) training to acquire the intellectual skills that will enable them to troubleshoot and devise better methods. Well-designed remuneration policies need to be offered that provide them with appropriate incentives to work hard to keep to the plan and to improve their skills. Without the appropriate HRM practices, workers could neither develop their intellectual skills nor feel motivated.

With a focus on workers' groups on the shop floor, this chapter discusses how they work both on a daily basis and at moments of takt time change (in the second section), and what kind of HRM practices have been instituted and how these practices function (in the third section). The fourth section points out some reasons as to why the transfer of the TPS is successful, and the chapter ends with a summary of the discussion.

Work organization in a trim line

A trim line is an assembly line where various parts are fitted into a car body, such as speakers and regulators into the inside of a door, for example. One glance at the assembly line gives the impression that workers are engaged in hard, monotonous labour without much time for rest, when in fact, flexible levelled production producing high-quality products is being conducted. It is even more difficult to observe the ways by which production is supported by workers' groups. This section aims to clarify this process as follows: first, it briefly describes the outline of the trim line of the most popular commercial vehicle in Indonesia, the Kijang.[3] Second, it examines the task profiles of the operators and the group leaders during daily production. Third, it discusses production control activities on the shop floor, such as daily quality meetings, an idea suggestion plan and quality control circles (QCCs). Fourth, it takes a closer look at how they cope with the change in takt time. Fifth, it describes their careers. Finally, it provides a summary of the findings.

The Kijang trim line is part of the final assembly section of the TAM assembly plant. The line is run by one supervisor, four foremen, twelve group leaders and eighty operators. Since summer 1998, the line has been assembling twenty-two types of the Kijang with takt time being set at 246 seconds. There are three basic types of the Kijang, and each of these has several subtypes with different engines, body sizes, transmission types and so on. The cycle time[4] differs between each of the three basic types: 260 seconds for the Grand Luxury (GL), 232 seconds for the Standard (STD) and 246 seconds for the Deluxe (DLX). With takt time set at 246 seconds, the production sequence is, for example, GL (260 seconds), STD (232 seconds) and DLX (246 seconds), so that no waiting time or line stops occur and levelled production is fully realized.

Operators are responsible for maintaining the quality standard and work according to the standard worksheets. Standard worksheets contain precise descriptions of all tasks related to a job, including the time necessary for walking and to complete each task. They also show the sequence of tasks, jigs and tools needed, as well as the location of stock. Figure 11.1 illustrates a standard worksheet.

No.	Task	Time (seconds) Operation	Walk
1	Watch *Harigami* (a standard manifest).	2	
	Pick up the tool and parts.	10	
	Walk to a body.		6
2	Install the five clamps in the wire harness.	30	
3	Put the wire harness in an engine room in order and cut the clamps.	15	
	Install the grommet of the turn signal.	5	
4	Install the two clips of the hood lock control cable.	10	
	Install the dumper fender into the hood.	5	
5	Shoot the bolts.	10	
6	Install the fuel filter and shoot the flange bolts.	15	
	Return to the right position.		4

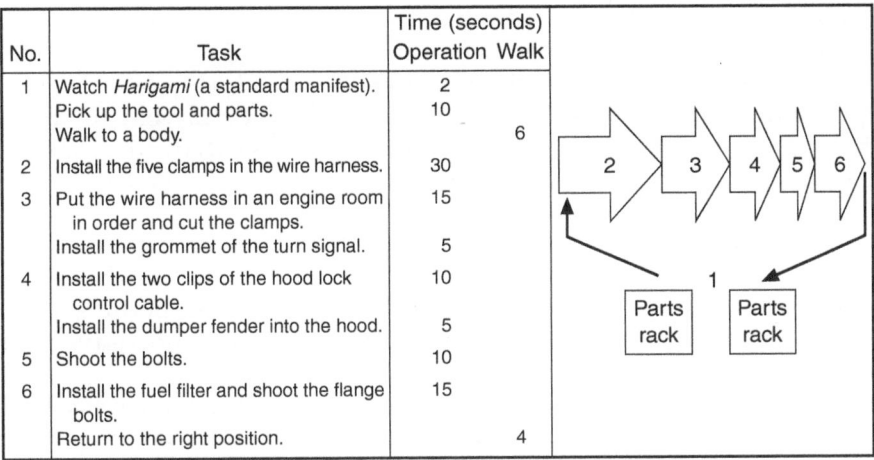

Figure 11.1 A standard worksheet

Source: Drawn up from Toyota-Astra Motor's company data 'A Standard Worksheet in the Trim Line' (modified)

Operators are expected to follow directions exactly as prescribed in a predetermined procedure, and in order to meet the quality standard, an operator has to adhere to the standard operation procedure, consisting of three sets of instructions: the standard manifest (*Harigami*), an assembly manual and a work ability sheet. The manifest indicates the type of Kijang being assembled,[5] and is put on the inside of a hood. The assembly manual outlines the sequence for the assembly of the parts. The work ability sheet describes the difficulty level of the operation, safety check points and the jigs and tools necessary for the job.

Although operators perform their jobs following the standard worksheets, they may not assemble the same parts in the same way on every job. Operators assemble twenty-two types of the Kijang, each of which requires different parts, assembly procedures and cycle times. In spite of the repetitive nature of the job, operators must pay close attention to the tasks at hand, since there is a wide variety of types.

No matter how carefully an operator assembles the parts, problems still occur. When a problem arises, an operator is expected to pull a line stop cord calling the group leader immediately. While the group leader is addressing the problem, the operator either performs other tasks or just watches the group leader fixing the problem. Operators are responsible for detecting problems as soon as possible, but fixing them is the responsibility of the group leader.

Operators at TAM are trained to be multifunctional through a system of job rotation. The Collective Agreement reads, 'For smooth work operation and the effective use of manpower and also developing capability (skill) of the worker, the company may do rotation within a department.'[6] The term 'multifunctional' in the trim line means that an operator is able to perform more than one assembly job. For example, a worker might be able both to assemble window regulators and to install side glass into doors. Figure 11.2 shows that approximately 20 per cent of the thirty-two operators in the trim line can perform two jobs, 15 per cent can perform three jobs and about 10 per cent can perform four jobs. These figures might represent an under-estimate of the percentage of multifunctional operators because the trim line was recently reorganized and the verifying test was not yet completed at the time this research was conducted.

A look at the chassis line, next to the trim line, reveals a higher ratio of multifunctional operators: as many as 45.5 per cent of twenty-two operators in the chassis line can perform two jobs, 13.6 per cent can perform three jobs, 9.1 per cent can perform four jobs and 31.8 per cent can perform more than five jobs.[7]

Multifunctional operators are one requirement for flexible levelled pro-duction and the TPS at TAM. To illustrate, takt time at TAM varies according to production levels. A change in takt time results in a reorganizing of each job. As reorganization applies to all jobs, changes in takt time are much easier to implement when operators are multifunctional, rendering any necessary reorganization of jobs a more efficient process. For example, suppose that takt time is two minutes and the number of operators is six persons before takt time change, and that takt time is shortened to 1.7 minutes in order to cope with an increase in production volume. The number of operators then increases by one person, from six to seven.[8] All seven jobs would have to be reorganized so that the cycle time would be 1.7 minutes. An important factor is the setting up of new jobs without hindering the smooth flow of production. One possible solution, then, for this scenario would be to pick the tasks requiring 0.3 minutes from each of the previous six jobs and bundle them all together to create the new seventh job.

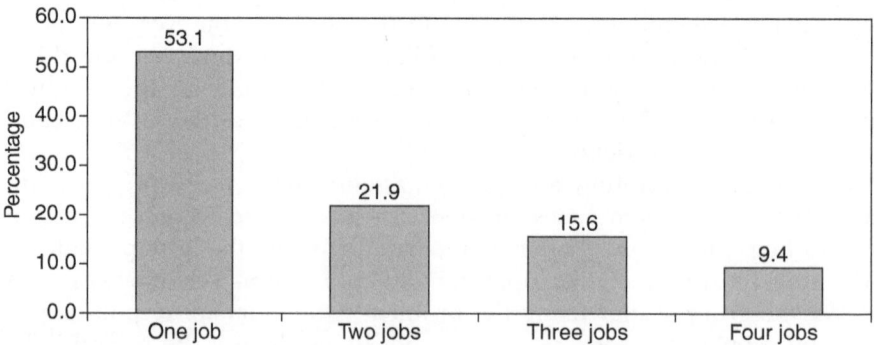

Figure 11.2 Percentage of mutlifunctional operators ($n = 32$)

Source: Derived from Toyota-Astra Motor's company data 'Multifunction Work in the Trimming Line'

Note
If an operator is evaluated and rated at more than 50 per cent, we can assume that the operator can perform the operation at 100 per cent for calculation purposes

However, this method would not result in a smooth production flow. Another possible way of reorganizing the jobs would be to take the tasks requiring 0.3 minutes from job 1 and move them to job 2, and transfer tasks requiring 0.6 minutes from job 2 to job 3 and so on,[9] as shown in Figure 11.3. Every operator, except for operator number 1, undertakes somewhat different job tasks than before. In a case where there are no multifunctional operators, everyone except for operator number 1 would need to be trained. However, when all operators are multifunctional, only one operator, the newly assigned operator number 7, receives training. By making the change in takt time much easier, having multifunctional operators increases the flexibility of the production line. Multifunctional operators thus represent one of the most important elements in attaining flexible levelled production and therefore the TPS.

With regard to group leaders, the trim line has twelve, each supervising seven to eight operators. Group leaders' daily responsibilities include performing multiple roles on the shop floor: they prepare for production, relieve operators, solve problems on the spot and engage in quality control and issues of labour management. When an operator needs relief from his or her work, a group leader takes over. Another group leader or a foreman then fills in for the group leader. Hence group leaders must be able to perform all the jobs under their supervision. With reference to the group leader's role in problem solving, upon discovering a problem during the assembly process, the operator is expected to pull a line stop cord to call the group leader, who in turn responds immediately to fix the problem without stopping the line. If the leader cannot fix the problem within the cycle time, the line must be stopped. Problems that can arise are, for example, when an operator assembles the wrong parts, forgets to install parts, or uses the wrong bolts or other fasteners, or when the quality of parts is unacceptable. A line stop will also be initiated when an operator has difficulty finishing a

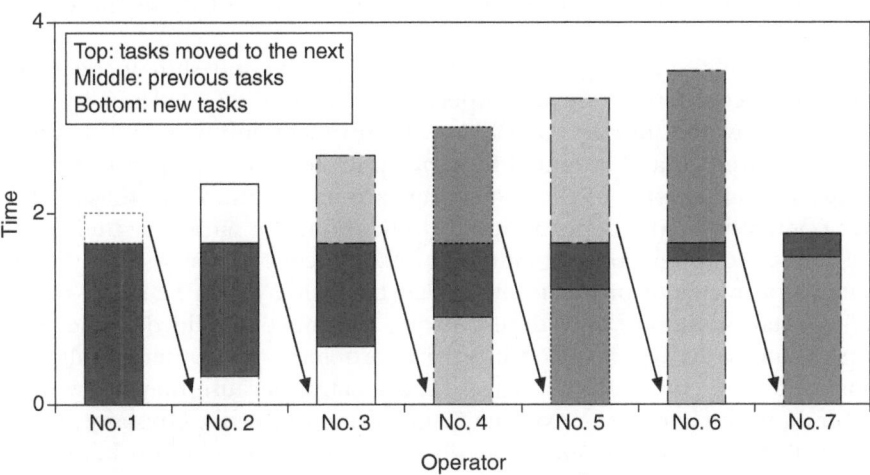

Figure 11.3 A decrease in takt time and job reorganization

particular job within the cycle time. The following two points are important concerning troubleshooting. First, problems do not always result in line stops. Group leaders endeavour to avoid line stops and limits are set for line stop time. For example, line stop limits may be set at twenty minutes per day per supervisor, seven minutes per day per foreman and two minutes per day per group leader. In this way, group leaders are encouraged to fix problems within the limits of the cycle time as often as possible. Second, group leaders have to be able to perform every job in their area, otherwise they could not fix problems effectively. While group leaders fix many problems by themselves, some initiate line stops and request assistance. In such instances, group leaders report to their foreman and request further instructions.

Every hour, group leaders go to the final inspection line to check whether defects have occurred in their areas. The inspectors in the final inspection line display tables showing the types and number of defects produced in each group leader's area. When the same defect is discovered in two or three consecutive units, the group leader is immediately informed. The group leader then consults with the operator responsible for the defect to determine the cause. Characteristic of these quality control activities is the speed with which group leaders are notified, but this process focuses on handling problems and defects that have already occurred.

Group leaders are responsible for keeping records of line stops and defects in order to 'build in quality in the production process'. While the above-mentioned activities constitute a portion of the quality control process, other quality control activities are also implemented. To ensure high quality, the occurrence of defects must be prevented. Group leaders keep records of line stops and defects. They usually draw graphs to illustrate the performance of their respective areas. The records and graphs are utilized in other quality-related activities.

Other than their daily work, workers at TAM are engaged in a number of activities in support of quality control: a daily quality meeting, the suggestion plan and quality control circles (QCCs). A discussion of these three types follows.

A daily quality meeting is held after each production shift to deal with problems encountered that day, especially issues that affected quality, cost, delivery, safety and morale (QCDSM). The thirty-minute meeting[10] is supervised by a foreman and attended by a group leader and the operators in the group. Several features of these meetings are important. First, the meeting gives operators a daily opportunity to think about the nature of their problems and to consider possible solutions. Second, concrete issues that directly relate to problems of production, such as balancing jobs, line stops, quality and new takt time, are considered during the meeting. Third, some operators are more actively involved in solving problems than others. Multifunctional operators, for example, may be more active because they know many jobs and tend to have more experience in solving problems. Operators spend almost all their time on repetitive, physically hard assembly work, and the thirty minutes they spend on problem solving may not appear to be enough time to make significant contributions to the smooth and efficient running of the assembly line. However, workers sometimes do come up with devices that render their processes foolproof so that problems do not recur.

The quality meetings, then, are closely tied into the idea suggestion plan. Two kinds of preventive measures are proposed at these daily quality meetings: the way jobs are performed (e.g. the location of parts and the jigs and tools used, and the adding of other instruments) either need or do not need revision. In the latter, more frequent case, the operator found to have caused problems is given advice on how to follow the standard worksheet and standard operation procedure more precisely. A proposal for revision is usually conceived and submitted in the form of a written suggestion. An operator who comes up with a concrete idea shares it with his foreman and asks for his opinion and advice. After the foreman's approval, the operator fills in a suggestion form in which the operator explains the problem, analyses its cost, depicts its current condition, suggests an improvement, explains a new method and/or new process and evaluates the result of the suggested improvement using data and cost analysis. Suggesting an idea does not seem an easy affair. It requires a different set of skills from those physical skills that are normally associated with production work. Workers must have intellectual capability to solve problems effectively and to submit suggestions that improve the process. In 1997, an operator or a group leader submitted an average of about eight suggestions per month.

Some problems recur time and again even though suggested solutions have already been implemented. The idea suggestion plan by an individual worker is brought to the QCC where it is discussed and QCC attempts are made to resolve these difficult issues. Almost all the workers participate in QCCs; in 1997, there were about 320 QCCs throughout the plants at TAM. QCCs are expected to meet for one and a half hours every Tuesday after the shift. In reality, how often QCCs meet is left to the respective members. Some QCCs convene every week, while others meet as little as once a month.

A QCC at TAM is usually composed of seven to ten members consisting of a circle leader, a theme leader, a facilitator and members. An engineer is not formally involved in a QCC and, in fact, a QCC seldom solicits an engineer's support. The QCC leader is a group leader and the facilitator is a foreman. A theme leader is an operator whose job is subject to the improvements being considered; thus the theme leaders are the ones who actually lead the QCC because they are the most knowledgeable about the situations being discussed. Other multifunctional operators, including a group leader, may also be active since they too are familiar with the problem. Between 1996 and 1997, five QCCs in the trim line actively tackled difficult problems and produced remarkable results.[11] In one case, they reduced the defect rate from 0.02 to 0.006 per cent; in three others it was reduced from 0.3 to 0.09 per cent, from 0.23 to 0.05 per cent, and from 0.042 to 0.005 per cent respectively. The fifth QC obtained a reduction in the number of broken parts per week from seventy-seven to two.

Thus workers in the trim line at TAM perform some production control tasks during daily quality meetings, the idea suggestion plan and QCCs. Through each of these activities the workers contribute to smooth and efficient production.

Takt time is usually changed several times a year at TAM, such as following a large shift in production volume and when a new model is introduced. Job reorganization is the most difficult part of the takt time change procedure and requires the most energy.

Job reorganization proceeds as follows. A group leader arranges all the job tasks under his supervision. The time necessary to complete each task is pre-determined with standard time data. Together with the group's foreman, the group leader combines tasks into jobs with the intention of ensuring a smooth production flow. The group leader also determines the sequence of assembling parts, so that the cycle times of each job are calibrated to the takt time. To accommodate the various types or models in the trim line, the group leader has to reorganize jobs many times. With some types of vehicles, the cycle time of each job may exceed the takt time, while with other types, the cycle time may fall below the takt time.

It is not easy to reorganize all jobs so that cycle time is equal to takt time (or takt time plus or minus seconds). In practice, some of the reorganized jobs take longer than the predetermined cycle time set during the trial stage. When it is discovered that the reorganization of some jobs is technically impossible, improvements become necessary and these are targeted on the layout of equipment, parts rack, jigs and tools, as well as the way the work is performed.

A group leader performs the reorganized jobs to see whether the cycle time is equal to takt time (plus or minus seconds). If the test succeeds, the group leader then draws up standard worksheets and submits them to the group's foreman. The foreman examines them and makes some modifications. The takt time change procedure is complete when the foreman approves the standard worksheets. After the full implementation of takt time change, the continued search for better quality and higher productivity begins anew.

Let us take a closer look at this process using the launch of the new Kijang at the beginning of 1997 as an example. During the initial stage of process design, the production engineering division calculated takt time at 2.9 minutes, a target, however, that proved difficult to attain. Thus production started at a takt time of 3.2 minutes instead. After launching production, takt time shortened from the initial 3.2 minutes to the planned 2.9 minutes, and then further reduced to 2.55 minutes. How was this accomplished? Takt time was shortened as a result of the continuous improvement activities that were conducted by operators, group leaders and foremen at the daily quality meetings, through idea suggestions, and QCCs. The ways of performing jobs were checked daily and some tasks in some jobs were moved to other jobs. Some jobs were modified when excessive walking to accomplish tasks was discovered and subsequently removed. Other jobs were modified by relocating parts racks closer to the assembly line or devising special drawers for jigs and tools. These small, incremental improvements took place while the operators were getting accustomed to their new jobs. The result was a 20.3 per cent reduction from the actual start-up takt time, which was also 12.1 per cent below the targeted takt time calculated by the production engineering division.

Careers

Group leaders and foremen play a critical role in smooth and efficient production. How are they selected? Group leaders are promoted from among the operators through fierce competition. As shown in Table 11.1, on average, a group leader in the trim line who finished high school worked as an operator at TAM for eleven years and has eight years' experience in the trim line before promotion. The corresponding figures for a group leader having completed junior high school are fourteen years and eleven years respectively. The competition for promotion to group leader is very fierce and based on job capability. Among the group leaders with a high school diploma, one was promoted to group leader within four years, while it took another twenty years. This suggests that promotion to group leader is not based simply on the length of service. Neither does it suggest, conversely, that the experience is not necessary. Experience is surely the most important requirement for promotion to group leader. It is, then, plausible to deduce that promotion is based on an evaluation of job capability within the context of fierce competition.

There are two routes to becoming a foreman. Foremen are selected either from the pool of group leaders using similar rigorous criteria, or from among graduates from academies, such as two-year technical colleges or technical junior colleges. Academy graduates are promoted to foreman after joining TAM and receiving one year of training. The two kinds of foremen differ from each other in terms of length of service, experience working in the trim line and age. Foremen promoted from within the organization outnumber the academy graduates.[12]

Table 11.1 Careers of the group leaders in the trim line at Toyota-Astra Motor, Indonesia

		Education	
		Junior high school	High school
Number of persons surveyed		3	10
Length of service (years)	The range	18–18	8–23
	The average	18.0	15.0
Length of service before promotion (years)	The range	10–17	4–20
	The average	14.0	11.1
Working experience in the trim line before promotion (years)	The range	9–15	0[a]–15
	The average	11.3	7.9

Source: From interviews with an Indonesian assistant manager, three supervisors and three foremen conducted on 24 April 1998

Note
a One group leader had no prior experience in the trim line before promotion. However, he worked for nineteen years on the repair line in final assembly where defects are discovered and repaired. Thus he was presumed to have enough knowledge about the trim line. Other than this one exception, the least experience that a group leader had in the trim line before promotion was four years

Flexible levelled production and work organization

Flexible levelled production assuring high quality is supported by a group of operators, internally promoted group leaders and foremen. While operators repeatedly assemble parts into a car body closely following the worksheets and standard operation procedure, and keeping to takt time during daily production, they also engage in production control tasks at the daily quality meeting and through the idea suggestion plan and QCCs led by group leaders and foremen. Group leaders are responsible for daily troubleshooting and daily quality control, and also play an important role when the takt time is changed and jobs are reorganized. Operators and group leaders, together with foremen, implement many kinds of improvements to ensure a smooth launch of the production with new takt time. Group leaders and foremen who play important roles are mainly those promoted from within the company through fierce competition. These workers can be called workers with intellectual skill.

Human resource management

This section examines HRM practices designed to recruit workers and provide incentives for them to work hard and improve their skills. Special attention is given to training and remuneration policies.

Training policies

TAM provides its workers with various training opportunities both on and off the job. The Collective Agreement states, 'Realizing that it is necessary to increase the capability of work as the requirements in increasing the productivity, the Company will keep making efforts to increase the ability, knowledge and skill of the workers through education and work training.'[13] The following examines introductory training for newly hired operators, training for operators, group leaders and foremen, and training at the TMC plants in Japan.

Newly hired operators, who are usually high school graduates, receive one week of classroom training during which they are given a general overview of TAM, TAM culture, the *kanban* system (later known as a Just-in-Time system), the suggestion system, the QCC and other relevant topics. After the classroom training, they are assigned to departments in the plant and given on-the-job instruction by group leaders for three months. This period is called OJT at TAM.

After the initial training, operators are expected to acquire skills through daily activities. Workers learn from the performance of their daily operations, through rotation, by using the suggestion plan, and QCCs. For example, operators receive daily OJT by group leaders concerning the proper use of the *kanban*. Foremen sometimes instruct operators on how to generate improvements. In addition, under the terms of the suggestion system, every operator is required to make at least one suggestion monthly.

These suggestions are examined by a group leader and a foreman. This process of making suggestions, as with QCCs, provides a training opportunity for operators. In 1997, a training course for operators was developed and delivered at TAM. This class increases the skill level of operators and follows a line similar to that at the TMC in Japan. Each job-specific skill, such as welding, assembly and painting, is taught during the training period.

While training programmes for foremen and group leaders are numerous and very substantial, the Compulsory Training and the TPS training are particularly important. This is because they give foremen and group leaders opportunities to systematize knowledge and experience, as well as to brush up on their skills. The Compulsory Training[14] takes three consecutive months. Participants receive two days of classroom training every two weeks for about two months. This is followed by a presentation that the participants make about their problem-solving experience during the last training session. With regard to the Compulsory Training, the following three points should be stressed. First, the TPS is taught, as are various production control techniques, such as the problem-solving process, the fishbone diagram, scatter diagram, histogram, control chart and Pareto diagram. While group leaders are considered to have familiarized themselves with these techniques through the suggestion plan and QCCs prior to their promotion, the training presents them with the opportunity to add to their knowledge and experience and increase their skill level. Second, the fact that instructors are TAM personnel, usually supervisors or foremen, suggests that TAM already has employees who fully understand production control techniques and the TPS. Without their having a deep understanding of these techniques, TAM could not utilize its personnel for teaching purposes. Furthermore, since the instructors possess a deep understanding of the plants, they may use concrete daily examples in the classes. This could further help participants to understand the subject matter of the training at a deeper level. Third, the techniques and concepts of the TPS are taught not only in lectures, but also on the shop floor. Participants are able to observe problem solving discussions and plant conditions, and also to make a final presentation.

The TPS training[15] involves 40 hours of classroom instruction and has two kinds of courses: one for foremen and supervisors and another for group leaders. The TPS teaches elements of the TPS such as the *kanban* system, Just-in-Time, standardization of work and *kaizen* (continuous improvements). These courses provide foremen and group leaders with an excellent opportunity to systematize their knowledge and experience. The classes are taught by instructors from TAM, usually supervisors or above, which indicates that TAM already has employees with a deep understanding of the TPS. The internally trained instructors can provide participants with a better opportunity to deepen their knowledge about the TPS.

Another important aspect of the training programme at TAM is the practical experience that some employees receive when they attend training at TMC in Japan. Between 1994 and 1996, from sixty to eighty employees were sent to TMC, where they received training for periods ranging between three months and one year, helping them improve their practical skills and abilities. Operators, group leaders and foremen sent to TMC work together with Japanese workers on the production lines for two to three months. This provides a unique opportunity to take a closer look at Japanese methods under the TPS.

Remuneration

The remuneration policies at TAM enable workers to rotate flexibly between jobs and actively cope with job reorganization. They additionally offer workers strong incentives for skill improvement and promotion.

The basic wages are determined by grades and steps within each grade. Every employee is ranked according to a certain grade and then assigned a step number. For example, an operator with three years of experience is placed in grade 3 and given step 15. The basic wages for this operator correspond to the wages set for this grade and step. Figure 11.4 shows the extent to which basic wages differ from grade to grade as well as between steps within grades.

An increase in step and upgrading, both of which result in a basic wage increase, are attained by annual performance evaluations or by promotions. The grade system at TAM has twelve grades, each of which has 120 steps. The lowest two grades are for janitors and clerical office workers, while the highest four grades are for managers or above. Newly hired operators, usually high school graduates, are placed into grade 3, step 1, whereafter they generally receive an increase in step every year. How many steps employees are given differs according to their performance evaluations. Those with the highest evaluations earn an addition of fourteen steps.[16] Those with the lowest evaluations earn no additional steps and remain on

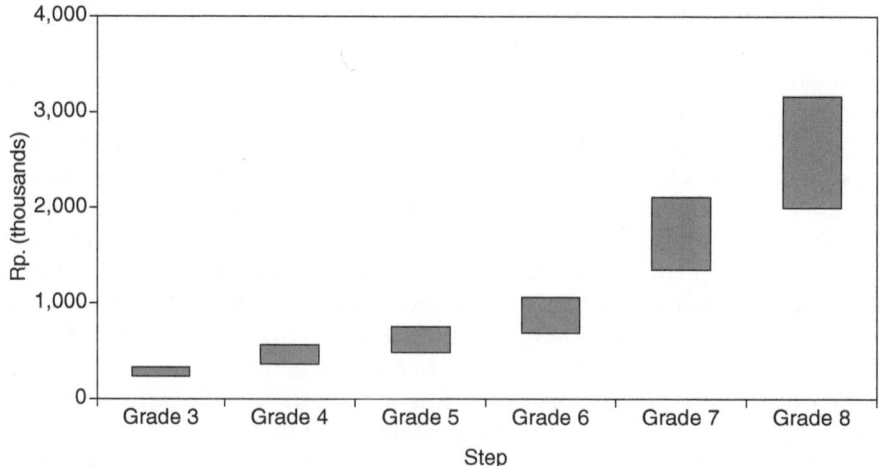

Figure 11.4 Basic wages

Source: Calculated from the interview record with an Indonesian assistant manager of the remuneration department of Human Resource Development, Toyota-Astra Motor, conducted on 13 March 1998

the same step. Thus employees receive an increase in step every year unless they receive the lowest evaluation. The increase in step is accompanied by an annual increase in basic wages, but the wage increase differs among the grades: the higher the grade, the larger the annual wage increase per step, as estimated in Figure 11.4. Hence upgrading brings employees higher wages.

Employees are moved to higher salary grades in two ways: by promotion and by the accumulation of length of service, provided that they earned good evaluations in the previous years. For example, group leaders promoted to foreman are upgraded from grade 5 to grade 6. When operators have worked for a certain period of time and received appropriate evaluations, they are upgraded from grade 3 to 4. With regard to grades beyond 5, promotion is the only route.

The characteristics of the basic wage system at TAM can be summarized as follows: first of all, the basic wages are not directly connected with jobs, and this type of wage system helps towards flexible reorganization of jobs when the takt time changes, and facilitates frequent job rotation. Second, the basic wages increase with length of service, and hence TAM's system can be considered to be seniority based. In this sense, workers are given an incentive to continue to work for TAM. Wages increase annually as a result of an employee's annual rise to a higher step, while they can also increase through placement to a higher grade. Even if employees are not promoted to a higher position, they can be moved to a higher grade within certain limits and thus receive higher wages. Third, because of the annual performance evaluation, the basic wages can differ between employees, even those with the same length of service and educational background. Fourth, the wage system provides workers with strong incentives to improve their job capability. If operators can prove that they can perform their jobs well and that they are capable of making positive improvements to their work, they enhance their chances of moving to a higher grade and receiving a promotion, which will lead to higher wages. The considerable difference in the basic wages between the grades strengthens the incentive and thus makes competition for promotion and upgrading fiercer. In conclusion, the remuneration policies strongly encourage workers to exercise flexibility and improve their job capabilities, while offering workers an incentive to remain in the company's service for longer.

Causes of successful transfer

The TPS has been successfully transplanted to TAM in Indonesia, and this section discusses the reasons why transfer was successful and the implications involved. First, it has been almost thirty years since TAM began its operation. It takes quite a long time to transfer know-how, especially management techniques. Second, TAM has developed one step at a time, starting its operation as an importer and distributor and gradually adding a manufacturing function. Steady development provided TAM with time to diffuse the technology successfully among Indonesian staff and workers. Third, TAM has had in its employ devoted Indonesian and Japanese personnel. Anecdotal evidence is given by Tomura (1997: 4–6) concerning Japanese staff, and also by Imai (1997: 126, 143–144) concerning Indonesian staff. However, the above-mentioned three conditions can be found among other Japanese multinationals, including companies that have not yet successfully transferred the management techniques. Thus these three conditions alone could not have brought success to TAM. Fourth, TAM has manuals for production system and production control techniques written in English and/or in Indonesian.[17] The manuals are important for at least the following two reasons: they enable the local staff and workers to understand easily the production management techniques that the TMC is attempting to transfer; and furthermore, they provide a management guide for the Japanese staff. It is often noted that the method of management varies depending on who is in charge of an overseas plant. The manuals may serve to decrease variations in the methods used. Nakamura *et al.* point out that 60–70 per cent of Japanese-related firms operating in Indonesia have manuals on production control such as quality control and 5S activities[18] in Indonesian or English (2001: 190–204), and that the preparation of various manuals significantly decreases the number of technicians dispatched from Japan to be stationed at the Indonesian firm (ibid.: 234–237).

Finally, the HRM practices have been developed internally by TAM's Indonesian staff by referring to those of the TMC and taking Indonesia's context into account. The development undergone by Indonesian staff seems to be relatively easy, since there is no substantial difference between the two countries in terms of either training or remuneration policies, as shown by Nakamura *et al.* (2001). Table 11.2 shows that some Indonesian-owned firms provide workers with training opportunities, although the ratio is much less than that in Japanese-related firms.

Table 11.3 shows that a quarter of Indonesian firms use a wage system for blue-collar workers that is based on ranking according to grade levels not necessarily closely related to jobs, and that the ratio is much smaller than that in Japanese-related firms.

According to Nakamura *et al.* (2001), 69 per cent of Indonesian firms have a seniority-based wage system for blue-collar workers, a percentage also

Table 11.2 Training being undertaken in Indonesian firms (multiple answer)

	Total (number)	No. training (%)	5S (%)	Technical skill up (%)	Quality control (%)	Supervisor (%)
Japanese-related	100.0 (100)	3.0	76.0	71.0	71.0	44.0
Indonesian	100.0 (200)	17.0	31.0	42.5	51.5	35.0
Total	100.0 (300)	12.3	46.0	52.0	58.0	38.0

Source: Nakamura *et al.* (2001: 101)

Note
Japanese-related firms are firms whose stock are owned by Japanese persons regardless of their share. Indonesian firms refer to the firms whose stock are not owned by Japanese persons, but by Indonesian persons

Table 11.3 The wage system for blue-collar workers in Indonesian firms

	Total (number)	Job rate (%)	Payment by results (%)	Position rate (%)	Rank rate (%)	Individually negotiated (%)	Others (%)
Japanese-related	100.0 (100)	16.0	2.0	28.0	45.0	3.0	6.0
Indonesian	100.0 (200)	26.0	7.5	25.0	26.5	6.0	9.0
Total	100.0 (300)	22.7	5.7	26.0	32.7	5.0	8.0

Source: Nakamura *et al.* (2001: 105)

reflected in Japanese-related firms (108), and 67.5 per cent of Indonesian firms have the merit system or performance evaluation system for blue-collar workers, compared with 86 per cent of Japanese-related firms making use of the merit system (111). In summation, Indonesian staff at TAM have not installed HRM practices that are totally different in nature from those already prevailing in Indonesia; hence, it is easier both for Indonesian staff to refer to the TMC's practices and for Indonesian workers to adjust themselves to TAM's practices.

Conclusion

The TPS has been successfully transplanted to TAM, Indonesia. Workers' groups on the shop floor largely contribute to the smooth and efficient running of the TPS. Various training programmes, including OJT, give opportunities for workers to increase their job capabilities, and the remuneration system provides workers with a strong incentive to improve on their job capabilities. These HRM practices, then, facilitate the technology transfer.

Important reasons for TAM's success are rooted in the facts that TAM has developed manuals about the production system and production control techniques written in English and/or Indonesian, and that the HRM practices have been developed internally by TAM's Indonesian staff by referring to those of TMC in Japan, while taking Indonesia's context into account.

In the meantime, TMC, like other Japanese multinationals, is planning to restructure its international division of labour in the Asian area. Simply put, it plans to produce important components and vehicles not only for each country market, but for the entire Asian market. The innovation that takes place in a transplant realized by multinationals, hence, will affect workplaces in other countries in the era of global competition.

Notes

1 See Nakamura and Wicaksono (1999) for more detail.
2 Toyota Astra Motor, 'Company Outlook of TAM', p. 8.
3 *Kijang* means 'dear' in English.
4 Cycle time is the time necessary for a worker to finish one cycle of his or her job with one automobile. The length of cycle time depends on how many tasks are included in a worker's job. Every worker completes his or her work on one vehicle at the same cycle time. Otherwise, production would not proceed consistently. Cycle time is

different from takt time. The latter means production line speed or conveyor speed. Under flexible levelled production, workers produce different types of products at different cycle times on the same line, with takt time being constant.

5 *Harigami* (the manifest) specifies such items as the type of Kijang, engine type, body type, whether the vehicle has an air conditioner, what type of audio system, cylinder type, how many doors, etc.

6 Collective Agreement 1995–1997, Article 19.1.

7 Source: Derived from TAM's company data, 'Multifunction Work in the Chassis Line'.

8 Six persons multiplied by two minutes, then divided by 1.7 minutes equals approximately seven persons.

9 In other words, all tasks in the line are distributed to seven jobs, following production flow, so that cycle time may be at 1.7 minutes.

10 Having a quality meeting daily may be a temporary expedient because of the severe market situation. Unfortunately, we do not have data that identify how often a quality meeting is normally held.

11 The result about the QCCs in the trim line is drawn from TAM's company data, 'A QCC Report in the Trim Line'.

12 Some foremen promoted internally are further promoted to supervisor. There is also another source for the pool of supervisors: university graduates are promoted to supervisor status after joining TAM and receiving one year of training. These two different kinds of supervisors differ in length of service, experience working on the final assembly lines and age. A greater number of supervisors are promoted internally following the path from operator to group leader, foreman and finally to supervisor.

13 Collective Agreement 1995–1997, Article 26.

14 The Compulsory Training was first introduced in 1995, by combining various training courses that had been already imported from TMC in Japan.

15 The TPS training was imported from TMC in Japan in 1993, with some modifications.

16 If an operator with grade 3, step 15 is evaluated as the best, he will be placed into grade 3, step 29 (= fifteen steps + fourteen steps).

17 For example, there are *The Toyota Production System* in English, a booklet titled *Toyota Production System* in Indonesian, manuals on roles and responsibilities of first-line supervisors written in Indonesian (Imai 1997), a booklet titled *Perlakuan Kondisi Kualitas yang Memburuk* (Troubleshooting of Defects) written in both Indonesian and Japanese, and *Petunjuk Ringkas Aktivitas Q.C.C.* (A guide to QCC activities) written in Indonesian.

18 The 5S activities refer to *seiri, seiso, seiketsu, seiton, shitsuke*, which translate to: tidiness, order, cleanliness, getting rid of anything useless and rigour.

References

Imai, M. (1997) *Gemba Kaizen: A Common Sense, Low Cost Approach to Management*, New York: McGraw-Hill.

Koike, K. (1999) *Shigoto no Keizai-gaku* (Economics on Work), 2nd edn, Tokyo: Toyo Keizai Shinposha.

Nakamura, K., Husodo, Z.A. and Hadiwijoyo, U.M. (2001) *Management Comparison and Localization: Indonesia and Japan*, Jakarta: Yayasan Obor Indonesia.

Nakamura, K. and Wicaksono, P. (1999) *Toyota in Indonesia : A Case Study on the Transfer of the TPS*, Jakarta: Center for Japanese Studies, University of Indonesia.

Tomura, K. (1997) 'Technological Transfer in Automobile Industry', unpublished seminar paper presented at the 'Technology Transfer and Mutual Cooperation between Indonesia and Japan' Conference, Center for Japanese Studies, University of Indonesia.

Unpublished internal company documents

TAM (1997) 'Company Outlook of TAM'.
TAM 'Collective Agreement 1995–1997'.
TAM (1998) Interview record with an Indonesian assistant manager of the remuneration department of HRM (conducted on 13 March).
TAM 'Multifunction Work in the Chassis Line'.
TAM 'Multifunction Work in the Trimming Line'.
TAM 'A QCC Report in the Trim Line'.
TAM 'A Standard Worksheet in the Trim Line'.

Personal communication

(1998) Interviews with an Indonesian assistant manager, three supervisors and three foremen, 24 April.

Part III

Industrial organization and skill formation in Europe – with Japanese influence?

12 Training and knowledge transfer in the European automotive industry

The impact of 'lean thinking'

Ben Dankbaar

This chapter offers some reflections on training and knowledge transfer in the European automotive industry. They are based on insights derived from a series of research and networking projects supported by the European Commission (EC), the European Automobile Manufacturers Association (ACEA) and the Association of Automotive Suppliers in Europe (CLEPA) in the 1990s and the beginning of the 2000s.[1] Training the workforce, at all levels, has become a matter of strategic importance in view of the manifold changes taking place in the industry. New skills are required to deal with these changes. On the one hand, there are new skills required in relation to the introduction of new product features (many of them related to the use of electronics), new materials (plastics, aluminium) and new levels of mechanization and automation. On the other hand, there are new skills required because of organizational changes taking place in the industry. The first type of skills can usually be acquired by traditional training measures, sometimes in combination with on-the-job learning. They are skills of the technical know-how type. Skills of the second type are mainly inter-personal in nature: skills in working together with other people, leadership skills, communicative skills, skills in cooperating across the boundaries both of organizations and of countries. This chapter is mainly concerned with training and knowledge transfer related to skills of the second type: training related to organizational changes in the industry.

The industry has undergone changes in both work and industrial organization. Changes in work organization include the introduction of teamwork, self-inspection, continuous improvement and flattening of organizations. Changes in the industrial organization include the creation of a supplier hierarchy with first-, second- and third-tier suppliers, simultaneous engineering, co-makership relations and generally higher levels of outsourcing on the part of car manufacturers. Many of these organizational changes were inspired by Japanese practices, or maybe it would be more correct to say that they were influenced by what were believed to be Japanese practices. Or, to make it even more complicated: they were influenced by what were believed to be organizational features of the Japanese automobile industry that were themselves believed to be the cause of the competitive successes of the Japanese industry in the 1980s. In 1990, a research team at the Massachusetts Institute of Technology (MIT) published an influential comparative study of the global automotive industry in which it was argued that the competitive successes of the Japanese industry could be explained by its highly efficient organization (Womack *et al.* 1990). In design and development as well as in manufacturing and assembly and even in distribution, the Japanese car manufacturers were offering new models of operating requiring much less labour input for each activity than was needed in the European or North American car industry. That is why the model was called 'lean production'. Since then, the Japanese car industry has been less successful, and several Japanese manufacturers are now controlled by North American and European competitors. I will not discuss here whether the analysis offered by the MIT study was correct or not (Dankbaar 1993; Sandberg 1995; Boyer *et al.* 1998). Obviously, the relationship between high productivity and international competitiveness is highly complex. A company may be highly productive compared to its international competitors, but may still lose the competition, e.g. because of an adverse currency exchange rate. Also, the description of Japanese practices in the MIT study is not very detailed and is open to debate. Nevertheless, what is important for our discussion here is that the discourse regarding the sources of Japanese successes experienced during the late 1980s and early 1990s has inspired organizational changes in the industry, both at the level of the workplace and at the level of industrial structure. Both these types of changes have prompted discussions about new skills that may be required of the workforce in order for these new structures to function properly. Before we enter into the debate concerning these requirements and how they can be fulfilled, some preliminary remarks on the importance and status of training in the industry are in order.

There is probably widespread agreement in the industry that modern manufacturing and assembly environments require higher levels of skill and intelligence than were needed in the past. On the one hand, more and new skills are needed because of the introduction of new product features, new materials and higher levels of mechanization and automation. Automation, in particular, has led to the elimination of many unskilled jobs, with the result that the average level of skill requirements for the remaining jobs is higher than before. On the other hand, one can argue that, certainly in assembly, the ability to carry out multiple jobs (multi-skilling) does not involve having a higher *level* of skill. Also, in the past, the abilities of most workers were probably grossly underutilized so that giving them more complex tasks would have involved some additional training, but not really have required higher levels of initial education.

There are very large differences between national industries in Europe and between factories within individual countries regarding the average level of formal initial training achieved by automobile workers before entering the factory. As a result, the training policies of companies will differ between countries and locations, depending on the levels of skill available in the different countries or regions. Most unions and employers are interested in making training a matter of government concern as much as possible. Unions expect this to result in more portable skills, enhancing workers' chances of mobility (and therefore give them more power on the labour market); employers see it as a possibility to socialize the costs of at least all non-company-specific training. In some countries, individual companies and employer organizations participate in apprenticeship systems and other national or regional training schemes, including an involvement in the definition of the qualifications that should be the result of these schemes.

Because of the differences between national institutions for education and training and the close links between these institutions and other elements of national culture, there appear to be only limited possibilities for the transfer of training policies from one country to another. In terms of the average initial level of training of automobile workers, the situation in Japan may be more comparable to that of Germany than to that of France or the United Kingdom in the sense that in both Japan and Germany, the average levels (number of years of initial training) are relatively high. However, in areas of high unemployment, new plants may also attract workers with high levels of initial training in France, Italy, or the United Kingdom. Specific to Germany is the impact of the apprenticeship system, which has resulted in a relatively large number of workers who have completed formal technical vocational training.

Against this background, it is obvious that the influence of the 'Japanese model' on training policies of automotive companies will be only indirect. There will be little or no imitation or downright copying of training measures. If there is an influence, it will be through the adoption or imitation of what are perceived to be important elements of work organization and industrial organization in Japan, which may have an impact on training requirements. In accordance with this line of thinking, this chapter's first section briefly explores some of the organizational changes that have taken place in the European automotive industry over the past decade. The second section reports on discussions concerning the training needs of the industry over that same period. The third section reports on research concerning training and knowledge transfer in the relations between car manufacturers and their suppliers. The fourth discusses the question as to whether there is such a thing as a European approach in training and human resource development (HRD). In the final section, some conclusions are drawn about developments in training practices and the role of the Japanese model.

Organizational changes in the European automotive industry

New industrial models ('lean production'), technological change (automation) and the increasingly global character of competition (globalization) have resulted in restructuring at all levels of the industry. Inside companies, manufacturing and assembly processes have been restructured to take account of higher levels of automation, higher levels of product variety, higher and more constant levels of quality and higher frequencies of model, product and process changes. Most of these changes involve giving people at lower levels of the organization more responsibility, often accompanied by the introduction of teamwork and reduction of the size of all kinds of so-called indirect functions (quality, logistics, maintenance, planning). Direct workers have to know and do more, a change that is usually not too difficult to realize. It is often more difficult to change the attitudes and working style of first-line management and support staff, who have to change from a directive mode (telling workers what to do) to a more supportive mode (helping the work teams to help themselves).

Between companies, there has been a tendency by car manufacturers to engage in more outsourcing. This tendency is not just inspired by the Japanese example, where the automobile industry developed into a tiered, pyramidal structure after the Second World War. It is also caused by technological changes. Some new technologies, notably in the field of electronics, fall outside the areas of traditional expertise of the car manufacturers. They have thus far refrained from acquiring this expertise by taking over relevant companies, if only because the technology is developing so quickly that it would be difficult to be sure that they were indeed acquiring all that is needed. Also, parts and (sub-)systems that used to be made by the car manufacturers themselves have become increasingly more complex and subject to rapid change, so that it is worthwhile to spread the associated costs out over a larger number of customers than just one. For example, practically all seat systems are currently being produced by a small number of large seat makers supplying all car manufacturers.

Increased outsourcing has had a series of almost inevitable consequences. In order to streamline operations and reduce the costs of contracting and logistics, car manufacturers have not just increased outsourcing, but also decreased the number of suppliers with which they have direct contacts and contracts. These so-called first-tier suppliers are expected to organize the supply chain further downwards and to transmit the desires of the car manufacturer concerning quality, reliability, etc. Because car manufacturers can be very demanding customers, some suppliers have quietly retreated to a 'second-tier' status. Others, however, have been fiercely competing for the status of first-tier supplier. In the course of that process, a series of mergers and acquisitions between suppliers has resulted in the rise of a number of large supplier companies with a strong technological basis, specialist knowledge and global presence. Indeed, in terms of size and market power, these global suppliers have become comparable to the car manufacturers themselves. It has even been suggested – with reference to the role of Intel in the PC industry – that in the future, customers may force car manufacturers to offer their cars with components of a specific supplier, whose brand might be as strong as that of the car itself.

Increased outsourcing has also encouraged thinking about the design and construction of the car itself! Large, first-tier suppliers are increasingly expected to supply complete systems, sub-systems, or modules (the difference being that systems fulfil a specific function, whereas modules can be built into the car in one piece – and may of course also fulfil one or more functions). Thinking of bumpers, doors, seats, dashboards, braking systems, fuel systems, etc. in terms of separate entities has influenced the organization of design and development activities. Specialized suppliers are increasingly involved in development work at an early stage so that their expertise can effectively be used before the design of other parts of the car sets unnecessary (and costly) limits to the design of the parts they are supplying. This is called concurrent or simultaneous engineering. As a result, along with their technological expertise, suppliers have increasingly been developing their own product development capabilities. They participate in product development teams involving personnel of the car manufacturer and one or more supplier companies. The use of modules in its turn is influencing the organization of the assembly process. The assembly line at the car manufacturer becomes shorter if suppliers deliver complete modules to the line that simply have to be fixed to the vehicle. Assembly of the module itself takes place at the supplier company, maybe in premises close to the final assembly line (in so-called supplier parks), or sometimes even in the car factory itself by personnel employed by the supplier. No wonder, then, that the expression 'co-makership' has been introduced to describe the new relationship between car manufacturers and their suppliers.

These structural changes in the automotive supply chain and the ensuing changes in roles for all parties involved obviously have important consequences for the skills and knowledge required in different places. Suppliers have to expand and continuously update technological knowledge in their field of expertise, but they also have to acquire knowledge in the area of assembly and supply chain management, which they did not need before. Moreover, they have to develop their capabilities for product development together with their understanding of the place of their product in relation to the rest of the vehicle. Car manufacturers have to learn to cooperate with suppliers in development, let go of certain responsibilities and accept that they no longer are the experts in certain areas, while they must still maintain the capability to judge the performance of the suppliers, transfer knowledge in the field of assembly and product development to their suppliers and concentrate on the integration of all supplier expertise into well-functioning and good-looking products.

Implications for training of the workforce

Against the background briefly sketched above, the industry has been discussing the implications for training of the workforce since the early 1990s. In 1994, a working party of the ACEA produced a document listing fields of personnel development projects with regard to organizational and technological change (Box 12.1). The document was drafted in support of a request by ACEA to the EC to help the European automotive industry in its restructuring efforts. The EC itself had produced a document titled 'Communication on the European Automobile Industry' early in 1994 (EC 1994). Against the background of restructuring going on in the industry after the second oil crisis, the Commission had made it clear that there would not be any special programmes in support of the automobile industry. However, existing European programmes supporting structural economic change could also be used by automotive companies. Such programmes could support collective (pre-competitive) R&D activities, as well as training.

Box 12.1 Fields of personnel development projects with regard to organizational and technological change

---

1   **Implementation of new work structures (production, administration and R&D)**
   - Changing roles of foremen and working group members due to new principles in the division of labour and enlarged competencies/tasks.
   - Changing conditions for career planning for all management tasks and levels as a consequence of new working structures.
   - Necessity for acquisition of new qualifications (e.g. key qualifications, social competence, network and process-oriented thinking, project management especially for employees).
   - Implementation of 'interface' management, i.e. cooperational relations with adjoining working teams, service functions (personnel in maintenance, planning and development).
   - At an early stage, coordination and cooperation prior to the introduction of modern techniques and the implementation of new products within existing working structures.
   - Implementation of modern forms of communication (team talks) with a view to process optimization and quality assurance.

2   **New learning methods on the job**
   - Assessment and optimization of methods leading to self-qualification (computer programs).
   - Use of multiplier systems to cope with large-scale qualification activities.
   - Use of moderation and communication techniques by employees during working and team process.
   - Flexible application of various training methods.
   - Implementation of advisory networks for the support of teams and foremen.
   - Concepts in the field of 'learning while working'.
   - Concepts in the field of informal learning processes (latitude in teamwork, learning in the course of work, cooperative learning).

3   **Implementation of continuous improvement processes**
   - Implementation of visualization systems (e.g. with regard to quality and work safety) as motivating factors for the continuous improvement processes.
   - Implementation of flexible systems to generate and handle ideas.
   - Implementation of variable bonus systems.
   - The continuous improvement process as the team communication process.

4   **Vocational training and retraining on the shop floor**
   - 'Customer'-oriented assessment of skill needs and organization of realization.
   - Implementation of qualification processes that are close to the production process.
   - Learning through job rotation.

5 **Development of cross-cultural management training and increasing language programmes**
   - Increase in foreign language competence of managers, employees and apprentices/trainees.
   - Teaching of and reflections on specific cultural outlooks, attitudes and ways of acting in the course of globalization (also the contextual use of terms).
   - Creation of language programmes in connection with intercultural projects.

6 **Cooperation between manufacturers and equipment suppliers in the field of training**
   - Training for co-makership (including possible transfer of knowledge and technology from assemblers to suppliers).
   - Training on the implementation and use of Just-in-Time systems, including the utilization of Electronic Data Interchange.
   - Training requirements for Just-in-Time assembly of components.
     Source: List produced by an ACEA working party in 1994

[M]easures in the field of training and retraining to adapt to industrial change and progress in production systems have the dual goal of enhancing the competitiveness of firms, whilst at the same time avoiding unemployment.... Training and retraining measures should focus inter alia on two areas which are keys to structural adjustment and competitiveness, and which are highly important to the EU car industry ...:

- *Training based on the introduction, use and development of new or improved production methods.*
  New technologies and, in particular, the progress of flexible automation have introduced new production concepts, transformed production structures and the nature and organization of work. The technological changes have led to the development of only a few totally new job configurations, but existing jobs are developing progressively towards polyvalence. Teamworking requires staff to be flexible, which can only be achieved if efforts in the field of ongoing training are increased and the willingness of the personnel to participate in such measures is raised. New organizational techniques, such as total quality control, flexible production systems and 'Just-In-Time', new information technologies, changing needs of the market and of society, particularly with regard to environmental protection, the use of robots and of new materials are examples of themes which could be covered in such training.
- *Training reflecting the need for SMEs [small and medium-sized enterprises] to adapt to new forms of cooperation with major companies, particularly with regard to subcontracting.*
  The multiplication of links between assemblers and subcontractors, the shortening of product life cycles and diversification leading to

small series production change the nature of qualification, logistics and management. In this context, measures should focus on intensifying the cooperation in training between companies with common interests or which share common characteristics. This should also foster improved logistics management, technology transfer and the transfer of professional experience between large companies and small companies which are often their suppliers.

(EC 1994, quoted by Dankbaar 1995: 86–87)

Within the framework of its programme on continuing education (FORCE), the EC supported the creation of a network in the automobile industry with the explicit purpose of setting up further projects and networks in the field of training. Indeed, this first network, which reported on its activities in September 1995 (Dankbaar 1995), became the starting point for a series of activities that are still going on.

The first network consisted of a group of six industry experts from academic institutions located in all the major car manufacturing countries of Europe, working under the close guidance of a Steering Group with almost thirty members representing the industry, trade unions and the European Commission. The list of training themes identified by the ACEA training party (see Box 12.1) became the starting point for intensive debate in several meetings of the Steering Group. The six ACEA themes were further prioritized into three main fields:

1   training for new work structures

    a   in production;
    b   in research, development and engineering;
    c   in the global corporation;

2   training for co-makership;
3   new methods and approaches for 'learning while working'.

Obviously, a wide variety of training projects can be carried out within these three fields. In its discussions, the Steering Group also identified a list of six high-priority 'sub-areas'. These six sub-areas can be considered the core themes of the three main fields of training identified above. They reflect the training needs as perceived by the industry in relation to the ongoing process of organizational innovation in the mid-1990s. They are:

- changing roles of foremen or middle management and working group members resulting from new principles in the division of labour and enlarged competencies/tasks;
- the continuous improvement process as the team communication process;
- early coordination and cooperation prior to the introduction of modern techniques and the implementation of new products within existing work structures;

- changing conditions for career planning for all management tasks and levels as a consequence of new globalized work structures;
- training for co-makership (including the possible transfer of knowledge and technology from assemblers to suppliers);
- concepts for 'learning while working'.

These priorities identified by the Steering Group formed the basis for interviews with industry representatives as well as for a series of case studies of exemplary training activities carried out by the six experts with the purpose of identifying possible future training projects and networking activities.

The majority of the training activities reviewed in the area of *training for new work structures* focused on training for continuous improvement and teamwork at different stages of organizational redesign in the companies. Lessons learned include the need for enhanced efforts in the areas of training team leaders, worker motivation and the presence of a corporate strategy in organizational design, as well as towards the value of medium-term planning and forecasting of occupational profiles, and the importance of communication skills for middle managers. Various practical tools and plans for implementing training in these areas had been developed.

The case studies related to *training for co-makership* illustrate quite different approaches to co-makership, in terms both of key concepts and of organizational approach. They demonstrate the key role of car manufacturers and first-tier suppliers as initiators of training across the supply chain as part of their outsourcing strategies. But through their associations, SME suppliers can also initiate bottom-up change by promoting training in the field of automotive project management. Of particular interest among such projects is the wide variety of approaches and models of cooperation between manufacturers.

The case studies dealing with *'learning while working'/on-the-job learning* demonstrate different concepts of the 'learning organization'. They highlight the rationale for transforming the training function into a process-oriented facility, namely, the need for open, flexible and modern training media, especially for basic training and the value of quantifying knowledge through learning audits and by 'systematizing' ways of working. There was great interest in the development of flexible and adaptable training products using multimedia applications. At the time, the Internet was still a thing of the future and no one mentioned the possibilities of using it for distributed learning processes.

The main product of the activities carried out by this first international automotive training network was a report with concrete proposals for projects that could be supported by the EC within the framework of its Leonardo da Vinci Programme (Dankbaar 1995). In the following years, several of these projects have indeed been realized, but the main result of this first network was indubitably that an agenda for training in the automotive industry had been discussed and established with the participation of representatives from major car and truck manufacturers, large and small supplier companies, national and European industry associations and, last but not least, representatives of the trade unions.

Many projects, network meetings and conferences on various aspects of training have followed since then. Within both ACEA and the association of supplier organizations, CLEPA working groups remained active in the field of training. Within CLEPA, the working group is still active today (2002) and has, among other things, been deeply involved in the development of an *Internet* infrastructure supporting training in small and medium-sized enterprises. Within ACEA, the working group on training was dissolved in the summer of 2000. The car manufacturers decided that this was not an appropriate activity for their European association, which is mainly intended to give them a presence in Brussels. They do not deny that it may be useful to communicate and cooperate in the field of training but, given their size and small number, they do not need the European association to take initiatives here. And indeed, European networking projects in the field of training have been initiated and supported by Daimler-Chrysler, Volkswagen, Ford and Volvo, among others.

The ACEA/CLEPA White Paper

One important accomplishment of the ACEA and CLEPA working groups on training should be mentioned. Late in 1999, on the threshold of the new century, ACEA and CLEPA published a common 'White Paper on Education, Training and Learning' (ACEA and CLEPA 1999). Anyone who has ever experienced involvement in drawing up statements representing the views of an association of companies will know how difficult it is to come to an agreement; the member companies may have widely diverging views on almost everything. Obviously, it is even more difficult to draw up a statement representing the views of two different industry associations, one of which (CLEPA) itself has associations as members. The White Paper, therefore, is a major achievement and deserves to be studied, because it is not likely that another one will be produced in the near future.

As might be expected, the White Paper emphasizes that governments have an important responsibility in creating and maintaining educational institutions that deliver a workforce with the required abilities:

> EU member state governments must ensure that every young person in Europe leaves the education and training system qualified and skilled and with a basic understanding of mathematical, scientific and technical disciplines, as well as transversal skills, or 'key competences' (social, interpersonal communication, methodological and analytical).
>
> (ACEA and CLEPA 1999: 4)

The paper then continues with the following observations concerning the impact of the 'lean production' debate:

> After some years of discussion, there is today broad agreement inside and outside the [automotive] industry, that modern management principles leading to higher levels of productivity, quality, innovation and

cost efficiency (commonly called 'lean management') are key to the industry's success.... [I]t is now widely recognised in the automotive industry that the mastery of technological change and the effective implementation of new technologies and work methods are largely dependent on flexible organisations, which require a multi-skilled and highly motivated personnel that is fully committed to 'lean' concepts and practices. From this follows, that in addition to the conventional 'hard' factors (e.g. rationalisation, automation, cost efficiency, technology), the 'soft' factor 'humanware' has (again) become a competitive issue of superior importance. This development brings the issue of education and training on top of the automotive industry's priority list. But the European automotive manufacturers have acquired the experience over the last years of radical adaptation, that it is not so much organisational, technological change and cost reduction that has created the most important problems in the change process, but the difficulty to change people's mindset, perceptions, attitudes and mentalities, irrespective of the individual person's hierarchical position in the organisation.... The automotive industry is quite aware of the fact that 'lean thinking' cannot be bought as 'technical hardware', and, on the contrary, that 'lean practices' must be learned by each individual in the organisation in a painstaking, continuous effort.

(ACEA and CLEPA 1999: 11)

In discussing future training priorities, to a large extent the White Paper follows the earlier statements mentioned above. Under the heading 'New working structures', it mentions training requirements connected with the formation of teams:

- training for team competence (such as solving conflicts, group dynamics);
- training for job flexibility within the team;
- training for team leader/speaker functions (for example, handling information flows and meetings).

Under the heading 'New forms of functional integration', it notes that traditional dividing lines between various jobs and specialist activities need to be reassessed with possible consequences for educational and training policies. A new balance needs to be struck between specialization and the integration of tasks within one function. A trend towards functional integration has been observed with regard to:

- skilled work and the traditional demarcation lines between skills;
- the traditional skilled and non-skilled tasks;
- between the non-skilled direct and indirect production tasks; and
- between skilled/non-skilled production tasks and white-collar tasks.

Noted under the heading 'Problem solving, improvement activities and increased responsibility' is that employees must learn to discover ways to manage better and continuously improve the manufacturing process. The corresponding training requirements are identified as:

- training in statistical methods (e.g. regarding Total Quality Management);
- training in brainstorming methods and in the presentation of results;
- training in planning methods, time and motion analysis and ergonomics to enable employees to participate in job design and efficiency improvement in their own work area;
- providing information about overall company policy, business environment, company structures, the process flow, the customers and the products in order to enable employees to develop a broader view on how their individual job fits into the overall picture.

The paper furthermore emphasizes 'learning for globalization'. In order to support the ambitions of companies wanting to become 'global players', new capabilities are required on the part of employees:

- methods of international communication;
- fluency in at least one foreign language; and
- experience of working in different cultures.

In view of the multilingual situation in Europe, it is notable that lack of both linguistic proficiency and knowledge of other cultures is considered the main obstacle to mobility and international cooperation.

The White Paper argues that the new 'lean' production environment does not necessarily require increased skills in terms of specialist technical capabilities, but it does require qualifications that were 'long ignored in the traditional manufacturing environment' (ACEA and CLEPA 1999: 18). These 'transversal' skills or 'key competences' encompass 'all those skills that permit people to change their occupation, improve their level of technical understanding, adapt to new or changing jobs, lay the foundations for intercultural competence and to transfer from one job to another' (ibid.). Consequently, these competencies refer to:

- social and methodical competencies;
- the self-motivated willingness to engage in lifelong learning;
- 'learning-to-learn' activities;
- problem solving;
- willingness to engage in continuous improvement processes;
- ability to take initiatives;
- abstract thinking;
- ability to work in teams.

The White Paper argues that these and other key competencies need to be acquired at an early stage and should somehow be made part of initial education: 'Primary school education through to the close of compulsory schooling must therefore much more than in the past provide young people with these new basic competencies that are required in their future professional life' (ACEA and CLEPA 1999: 19).

Thus, the White Paper provides a wide-ranging, but sometimes quite detailed, overview of the training needs of the automotive industry in the wake of the debate on lean production. It closes with several proposals for action, which need not concern us here, with the exception of the proposal to create active 'European learning clusters,' because this proposal provides additional insight into the training needs identified by the industry. The main idea of the 'cluster' concept is the creation of partnership networks between different cultures of education in the European Union, bringing together educational players, companies, the social partners and local authorities and associations. Information and mutual recognition should be the guiding principles for these clusters.

Finally, the White Paper also devotes a section to the use of new information technologies in training. It is clear that, since the statements made five years earlier, the possibilities of using information technology in training have become greater. Indeed, the paper notes that the multitude of options available and the 'leapfrog' developments in this area make some consolidation and orchestration necessary. The high speed of developments and the absence of clear standards obviously make it difficult for companies to make choices and decide upon specific uses of technology.

Training and knowledge transfer between car manufacturers and suppliers

Within the framework of one of the EU-funded Leonardo da Vinci Programmes – 'The Training Network of the European Automotive Supply Industry' – the Nijmegen School of Management has carried out an investigation on behalf of the project coordinator, CLEPA. The purpose of this study was to gain a more precise insight into the new demands of car manufacturers regarding their suppliers and the implications of these new requirements for training activities and knowledge transfer between automobile manufacturers and suppliers. A report on the study was made public by CLEPA in the spring of 1999 (Dankbaar and Bouwman 1998; see also Bouwman 1998).

The investigation distinguished various forms of training and skill acquisition for automotive suppliers. It looked at (1) the size and contents of regular training activities in supplier companies; (2) the relevance and contents of the so-called supplier development programmes that are being offered by car manufacturers; and (3) the importance of co-makership relations for the acquisition of new knowledge and skills. The investigation was carried out by means of a brief questionnaire sent to all corporate members of CLEPA.[2] The questionnaire was developed in close cooperation with the Working Group Training of CLEPA. Apart from the questionnaire survey, a series of interviews were held with persons responsible for supplier development programmes at Renault, Rover, BMW, Nedcar (which manufactures cars for Volvo and Mitsubishi in the Netherlands) and MCC (manufacturer of Smart in Hambach, France), and with persons responsible for training and human resource management at automotive suppliers Robert Bosch (Germany), Magnetti Marelli (Italy), Lucas Varity (the United Kingdom) and Labinal (France).

The total number of respondents to the questionnaire was fifty-one, the majority of them large 'first-tier' suppliers of (sub-)systems. This is obviously not a representative sample of the automotive supply industry. It can be assumed that the demands made upon big first-tier suppliers will be higher than those on second-tier suppliers delivering components, machined or pressed parts. It is clear, however, that first-tier suppliers will, in time, have to make similar demands upon their own suppliers. Consequently, this investigation shows the direction in which demands on all suppliers are moving.

Training priorities

Over the previous five years, these companies had spent an average of 2–3 per cent of their annual turnover on training. Further questions concerning the contents of training made a distinction between three categories of personnel: production personnel, technical staff and first-level management (supervision). The respondents could indicate the relative importance of the following reasons for *training production personnel*: new organizational concepts (teamwork); technological changes; quality programmes; health and safety issues; and environmental issues. Each of these reasons could be given a score between 1 (not important) and 5 (very important). Quality programmes were indicated as the most important reason for training. Almost 55 per cent of the respondents considered quality programmes to be a *very* important reason for training. New organizational concepts and technological changes were also seen as very important by 47 and 42 per cent of the respondents respectively. Lower priority was given to health and safety, and environmental issues, although these items were also considered important to very important (scores 3–5) by the large majority of respondents.

The respondents could also indicate the relative importance of the following topics of training for production personnel: interpersonal skills (communication and leadership); problem-solving techniques; knowledge of foreign languages; statistical process control (quality control); general technical skills; specific technical skills; information technology skills; and production methods. Based upon the opinions, an overall score was calculated, which could vary between 100 (not important) and 500 (very important). The scores for each topic resulted in the ranking presented in Table 12.1.

The high ranking for 'interpersonal skills' appears to indicate that, next to technical skills, so-called human skills have become increasingly important, even for production personnel. Specific technical skills had a higher overall score, but interpersonal skills were most frequently indicated as 'very important' (32 per cent). Nevertheless, it is clear that most skills mentioned here are relatively well codified and can be transferred by means of traditional training programmes, either in a classroom environment or in on-the-job training (OJT).

Table 12.1 Relative importance of topics for training of production personnel ($N = 48$)

Ranking	Topic	Score
1	Specific technical skills	398
2	Interpersonal skills	392
3	Problem-solving techniques	382
4	Information technology skills	381
5	Production methods	379
6	General technical skills	373
7	Statistical process control	366
8	Foreign languages	327

Besides the relative importance of topics of training, the respondents were asked in an open question to mention the greatest training needs for production personnel in the future. The various topics mentioned were then consolidated into a smaller number of categories plus the category 'other', which comprises all items that did not fit elsewhere and were mentioned less than three times. Table 12.2 shows that the biggest training needs for production personnel were expected in the area of quality, followed by training in production methods (including Just-in-Time logistics) and teamwork.

Furthermore, respondents were also asked in an open question to provide important topics of *training for technical staff and first-line management* (supervisors). For technical staff, training in the field of quality was by far the most frequently mentioned (SPC, TQM). CAD, information technology and project management were also mentioned relatively frequently. Obviously, various technical skills remained important topics of training for technical personnel. With the exception of project management, organizational and/or interpersonal human skills figured less prominently than for production personnel and for first-level management (see below). In the future, further training needs of technical staff were expected in the areas of quality (still by far the highest score), specific technical skills and problem solving.

Table 12.2 Biggest training needs expected for production personnel in the future

Topics for training	Times mentioned	Percentage
Technical skills	6	9.5
Health and safety	6	9.5
Interpersonal skills	4	6.3
Production methods (JIT)	9	14.3
Leadership	3	4.8
Multi-skilling	3	4.8
Problem solving	4	6.3
Quality	14	22.2
Teamwork	8	12.7
Total productive maintenance	3	4.8
Other	3	4.8
Total	63	100

The training of first-level management appears to be most frequently concerned with leadership training (coaching competencies), interpersonal skills, teamwork and communication, although quality also remains an important item as regards this category of personnel. Another topic mentioned relatively frequently was 'problem-solving techniques'. Future training needs were expected to remain most urgent in the area of leadership skills, interpersonal skills and communication. Indeed, compared with current topics of training, the respondents clearly expected so-called human skills to increase in importance for first-level managers.

Knowledge can be transferred through different means. The respondents were asked to rank six methods for training production personnel. Experienced workers, including supervisors, providing OJT are most frequently used as trainers for other production personnel. This is often just part of their work, but sometimes experienced workers are assigned temporarily, but full time, as trainers. Around 70 per cent of the respondents ranked these two methods (OJT by experienced workers, full-time or part-time) in first or second place, after which they ranked outside consultants and full-time company-employed trainers. Least frequently used are trainers from regular professional training institutes (schools) and suppliers of equipment and materials, although still about one-third of the respondents placed these two methods first or second. In fact, outsourcing for other than OJT has become quite common. In several cases (Ford, VW, Renault and Fiat), the training departments of manufacturers have become separate organizations, which also offer their services to outside customers.

In an open question, respondents were asked to indicate what they considered to be the biggest problem in the field of training. Not surprisingly, the problem most frequently mentioned was lack of time. In the Just-in-Time manufacturing environment, time for training is scarce and pressures of production constantly dominate the agenda. Most production managers tend to bend under this pressure and give lower priority to training. Only if training is vigorously supported by top management can it be adequately organized in this environment. The second biggest problem mentioned is the lack of good evaluation and auditing tools in this area. Advocates of training need better proof of its effectiveness and a better means to evaluate individual programmes and trainers. A third problem mentioned was the availability of good training programmes. Finally, some respondents considered a lack of finances to be their biggest problem in the area of training.

Training in the supply chain

Training can also be provided by customers. Suppliers can participate within so-called supplier development programmes. Of the respondents, 35 per cent had experience with one or more programmes. The same percentage of suppliers had developed training for their own suppliers (the ones who developed such programmes are not necessarily the same as the ones who received training, although this is most often the case). The main focus of the supplier development programmes of car manufacturers is on quality

(80 per cent). The other topic of a supplier development programme is related to product development (design). The suggested other topics of training in the questionnaire – logistics and maintenance, and management skills – were not ticked at all. On average, these supplier development programmes were considered very useful. The respondents had experience with supplier development programmes provided by Nissan, Toyota, GM/Opel, Fiat, Ford, Renault, Rover, Volkswagen and BMW. Several had been involved in supplier development programmes offered by different manufacturers.

Out of fifty-one respondents, thirty-nine (76 per cent) indicated that their company had been given new responsibilities in the field of product development by car manufacturers in recent years. In 34 per cent of these cases, this led to the participation of one or more employees of the supplier in training programmes of the car manufacturer (related to product development and design). In 46 per cent of the cases, employees of the supplier were co-located at the car manufacturer for some extended period of time. This supports the assumption that knowledge that is not formalized and for it to be transferred or shared requires physical presence and direct contact between people. Almost all of these 'co-makers' (87 per cent) indicated that an important transfer of knowledge was or is taking place within the co-maker relationship. Differentiated for various areas of knowledge, the picture that emerges is shown in Table 12.3.

According to the suppliers who considered themselves co-makers, 62 per cent perceived a transfer of knowledge according to the needs of final customers from the car manufacturer to the supplier and only 12 per cent perceived a reverse knowledge transfer from the supplier to the car manufacturer. With respect to the knowledge of product design, there are flows of knowledge in both directions: 53 per cent of the respondents cited a knowledge transfer from the supplier to the car manufacturer, but also in 50 per cent of the cases, a knowledge transfer from the original equipment manufacturer (OEM) to the supplier was noted. Knowledge transfer regarding product technology was mentioned almost twice as much from the supplier to the car manufacturer (56 per cent) as from the car manufacturer to the supplier (29 per cent). This seems to indicate that car manufacturers depend on their suppliers for technical knowledge. Finally, in 47 per cent of the cases, knowledge about project management was transferred from the car manufacturer to the supplier, and in 24 per cent of the cases there was a knowledge transfer in the other direction. The increasing role of co-makers as organizers of design, production and assembly of complete sub-systems or

Table 12.3 Knowledge transfer between original equipment manufacturer (OEM) and supplier ($N = 38$)

Knowledge related to:	OEM → supplier (%)	Supplier → OEM (%)
Needs of final customers	62	12
Product design	50	53
Product technology	29	56
Project management	47	24

modules has confronted them with new demands in the field of project management, coordinating a large number of (sub-)suppliers and understanding their specific products and technologies. In view of this, it seems logical that there should be an important flow of knowledge about project management from car manufacturers to their suppliers.

To sum up, quality remains the most important topic for formal training at automotive suppliers, as well as in the supplier development programmes offered by car manufacturers. To the extent that quality is achieved by means of new organizational arrangements, training is focused on interpersonal and other human skills, as opposed to technical skills, which continue to constitute the bulk of most training programmes. Where relations between supplier and manufacturer have progressed most towards co-makership, new skills and knowledge are needed, which often cannot be transferred by regular training or development programmes, at least not at the moment, but require co-location and cooperation between employees of both enterprises.

Owing to the strict requirements imposed on suppliers by car manufactures, the way the two organizations communicate and cooperate appears to be changing dramatically, at least at the level of first-tier suppliers. The Bow-Tie versus Diamond approach, developed by Cooper *et al.* (1997), illustrates how the relationship between two companies changes (see Figure 12.1). The traditional interaction between the two companies is via a purchaser and a salesperson (the Bow Tie approach). All information is transferred through these two persons or departments. If we rotate these triangles until both sides touch each other, we have a new type of relationship. This is what Cooper calls the Diamond Approach. All functions can communicate across the boundaries of the two organizations.

Some of the newly required skills are relatively well defined and can easily (albeit at a sometimes substantial cost) be acquired by traditional training methods. This is often true for technical skills involving the mastery of new technologies. Others are less well described, being based on experience more than on explicit insights (experiential or tacit knowledge), and need to be transferred through some organized form of exercises or learning-by-doing,

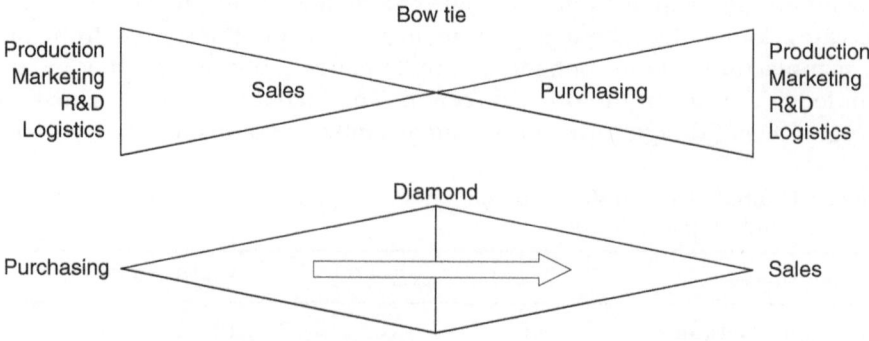

Figure 12.1 Bow tie versus diamond

Source: Adapted from Cooper *et al.* (1997)

on-the-job learning. Finally, there is experiential and/or firm-specific knowledge, which is not well enough defined to be made part of a training programme. This kind of knowledge can be acquired, and indeed discovered, only in actual practice. If it needs to be transferred to other companies, there is virtually no other way of knowledge transfer than to place the people concerned together in one spot and let them cooperate and communicate.

Are there European approaches in training and organizational change?

One could argue that a European car industry does not really exist. Although the mutual penetration of their national markets has greatly increased over the past decades, the European automakers are still first and foremost German, Italian, French, or British. History, years of investment, organizational culture, human resources and, last but not least, marketing image are all tied to the national roots of these enterprises and brands. The diversity of identities and images generated by European culture can indeed be considered an important source of competitive strength of European enterprises. However, there are also drawbacks to be considered. Probably the most important one is scale. European national markets are simply too small to serve as a platform for a participant in global competition. That is where the Japanese and Americans have an advantage: their national markets are much larger. It is one of the guiding ideas in the creation of the European Union that the European market would thereby acquire a similar scale.

In the areas of training and human resource management, national as well as local traditions and institutions in industrial relations, education, organizational cultures and management styles play an important role. The 'lean' organizational concepts launched in the 1990s had obviously been developed in a quite specific cultural and institutional setting (although it should be noted that in the process of describing Japanese practices in the English language, some of the 'Japanese' concepts were already given a decidedly Westernized interpretation and content that did not necessarily match the actual practices in Japan). The role of the trade unions, legislation on health and safety standards, working hours, systems of personnel evaluation and compensation, terms of employment (lifelong or temporary), the pension system, social security, labour market mobility, the role of women and the family and the organization of education are all very different in Japan from what is common in most of Europe. Many of these Japanese arrangements are not found particularly attractive by European workers, or indeed by their employers. That prompts the question: can organizational concepts be detached from the context in which they originated and successfully applied in another context?

It seems an odd question to ask, considering the fact that Japanese concepts and approaches were supposedly implemented right across the industry during the 1990s. The MIT study mentioned earlier brushed aside all doubts about the applicability of Japanese concepts in other countries by pointing to the success of so-called Japanese transplants (subsidiaries of

Japanese manufacturers) in the United States and United Kingdom. Why raise such theoretical questions, then, if practice has already moved beyond them? There are indeed several reasons why it is appropriate to do so:

- The fact that Japanese transplants are successful may bear no relation to their human resource management practices, but may stem from other factors, which may be human resource-related (like operating with a young workforce in a high-unemployment region) or not (like having good products, designed for easy assembly).
- Simply copying Japanese practices in human resource management will, if it is possible at all, seldom generate competitive advantages. It is highly unlikely that the effect on the performance of enterprises will be higher than in the context in which they originated, and usually it will be lower.
- Management at Japanese transplants may have adapted the practices of the Japanese mother company to the locally prevailing institutions and attitudes. In fact, they usually have done so and more may be learned from these 'hybrid' practices than from the Japanese original.
- The use of Japanese organizational concepts may be possible without much of a problem in the short term but, as the longer-term consequences become clear, it may result in serious social conflict. This may, for instance, be the case if new organizational forms offer fewer opportunities for promotion, which can result in discontent among people who would have gained promotion in the old organization.
- The use of Japanese concepts may consciously be used by management to introduce or even force through changes in the institutional environment in Europe, if they consider some form of 'Japanization' desirable. Industrial relations practices in Japanese transplants, for instance, have been an impulse for change in both the United Kingdom and the United States.

There are good reasons, therefore, to be aware of the interactions between the use of new organizational concepts in human resource management and the institutional environment. Training and motivating European personnel for world-class performance requires training programmes and organizational models that take account of the high performance standards set by the leading enterprises of the industry, but also of the prevailing institutional ramifications and the high expectations of European employees concerning the quality of work offered to them.

As noted above, European approaches are, in reality, often just national approaches and traditions. However, there are some values, traditions and tendencies that are common to many different national European cultures. A few where clear differences appear to exist with prevailing attitudes and arrangements in Japan are worth mentioning, such as equal treatment for men and women, equal pay for the same work, high appreciation of family life and of free time spent with family and friends and mobility between jobs and enterprises on the basis of universally recognized qualifications. Efforts to connect these 'European values' to principles of organizational design and human resource management usually come up with references to European traditions in socio-technical design and the important role of industrial relations and social dialogue.

- The European car industry can draw upon a long tradition of experimentation, as well as regular practice with socio-technical job redesign in production work. Long before Japanese concepts became popular, these European experiments already departed from the traditional forms of work organization that were introduced by Henry Ford in the early decades of the twentieth century. Especially in Sweden and Germany, experiments with a reduction of the division of labour (teamwork) and movements away from short-cycled assembly work have been undertaken since the mid-1970s. Under the influence of Japanese practices, the more radical movement away from the traditional assembly line was stopped and actually reversed in the second half of the 1990s. At the same time, however, these earlier experiments in work organization have coloured the reception and implementation of Japanese concepts. In the debate about the content and meaning of 'lean' concepts, the socio-technical tradition itself developed further into a more comprehensive approach of organizational redesign.
- In Europe, the emphasis on training and the importance of human capital for competitiveness almost automatically implies a call to pay adequate attention to the social dialogue between employers and trade unions. There are major differences between member states in legislation and other institutional arrangements concerning the involvement of trade unions, works councils and other representations of employees in matters of training and organizational and technological change. Therefore, the forms of social dialogue may differ according to subject and location. The need to have a social dialogue on these issues, however, has been repeatedly confirmed by all parties concerned in the European automobile industry. In numerous cases, this has resulted in contractual arrangements that have provided a solid basis to training activities.

Although it is extremely difficult to expound truly European principles in the field of training and organizational change, we may venture the following with special regard to the automotive industry:

- *Trade unions and workers have a clear preference for training resulting in qualifications that are recognized by all relevant employers.* The increasing importance of OJT and training while working would find more support if it resulted in such 'portable' qualifications. There is no reason why employers, unions and vocational training institutions could not come to an agreement concerning standards and certification of qualifications acquired on the job over the years. An appropriate recognition of the effective utilization of 'key qualifications' would also be worthwhile from this perspective. The preference for transportable qualifications also indicates the continued significance of vocational training institutions at the level of initial training.
- *The workforce and/or its representatives have a right to be consulted if major organizational change is being planned.* Whatever the specific traditions and

legislation in the various European countries, there is a deeply rooted notion of industrial democracy, which says that it is right and natural to consider the concerns and ideas of the workforce if important decisions are to be made about the organization of work. In the automotive industry, agreements about new forms of work organization have been concluded between management and works councils or other representative bodies in many enterprises and plants. Even though management has often been obliged to enter into such agreements on the basis of legal regulations, there is a broad consensus that, in most cases, such agreements have been helpful in achieving the aims of the enterprise.

Conclusions

By way of conclusion and summary, the following points can be made:

- The context of training is quite different in different European countries, and therefore the impact of new organizational concepts on training is also different across countries and indeed also between regions and locations. Nevertheless, a general trend can be noted towards higher average levels of skills in relation to the introduction of teamwork (multiskilling) and, more generally, in relation to the use of new technologies and the higher complexity of products.
- There are some general European approaches to dealing with new training requirements, which are characterized by the desire to create portable skills and a willingness to have a constructive dialogue concerning training and organizational change between employers and organized representatives of workers.
- The main themes of training programmes have evolved very little over the past decade. The themes mentioned in the ACEA/CLEPA White Paper do not differ very much from those listed by the ACEA working party in the early 1990s. There is a clear trend that interpersonal skills and 'key competencies' are becoming more important for production personnel as well as management. Renewing and updating technical skills remains, of course, an important activity.
- On the whole, co-makership relations are developing relatively slowly (Jansen and Peters 2001). This is reflected in the fact that training activities in the supply line are still mainly concerned with quality issues and not with product development (knowledge transfer, project management), although the latter themes are clearly becoming visible.
- With increased outsourcing, one would expect car manufacturers to lose expertise concerning the manufacturing and design of parts they used to make themselves. This may indeed be the case, but it does not seem to take the form of a direct transfer of technical expertise from car manufacturers to their suppliers. Suppliers do receive knowledge from car manufacturers, but mainly in the field of project management, quality and assembly, and not (or much less) concerning the products and processes in which they have specialized themselves.

- Where co-makership is indeed developing, co-location of workers from the supplier and the car manufacturer is frequently taking place for the duration of development projects. Furthermore, the creation of supplier parks and the assembly of components by supplier personnel next to (or even on) the car manufacturer's assembly line have created conditions for an optimal exchange of knowledge and experience.

Notes

1 The author was involved in the following EU-supported projects: The CVT/FORCE Automobile Network (1994–1995), under the FORCE Program; the ACEA Learning Network (1996–1998) and the Training Network for the European Automotive Supply Industry (1996–1998); and Automotive Learning and Experience Transfer (ALERT, 1997–1998), the ELAN Benchmarking Initiative 2000 (1999–2001) and ELAN 2 (2001–2003), all under the Leonardo da Vinci Programme. The Training Working Group of CLEPA, of which he is an associate member, was involved with the Esprit project Total Quality On-Line (1999–2000), which created a virtual learning environment for automotive suppliers, and its successor project E-Learning Solutions for Automotive Supply SMEs (ELSA) (2001–2003). The first two network projects mentioned each published four newsletters on training issues under the titles *The ACEA Learning Network* (1996–1998) and *Skills and Suppliers* (1998–1999).
2 The membership of CLEPA consists of approximately fifty corporate members (mainly large corporations) and the national automotive supplier associations of car-manufacturing European countries. Efforts to have the questionnaire distributed by the national associations among their members were unsuccessful.

References

ACEA and CLEPA (1999) *1999 ACEA/CLEPA White Paper on Education, Training and Learning*, Brussels: ACEA and CLEPA.

Bouwman, B. (1998) 'Supplier Development: A Set of Requirements', MA thesis, University of Nijmegen.

Boyer, R., Charron, E., Jürgens, U. and Tolliday, S. (eds) *Between Imitation and Innovation: The Transfer and Hybridization of Productive Models in the International Automobile Industry*, Oxford: Oxford University Press.

Cooper, M.C., Ellram, L.M., Gardner, J.T. and Hanks, A.M. (1997) 'Meshing Multiple Alliances', *Journal of Business Logistics*, 18 (1): 67–89.

Dankbaar, B. (1993) *Economic Crisis and Institutional Change: The Crisis of Fordism from the Perspective of the Automobile Industry*, Maastricht: UPM.

—— (1995) *Learning to Meet the Global Challenge; A Contribution to the Debate on Continuing Vocational Training in the Automotive Industry of the CVT/FORCE Automobile Network, 1994–1995*, Maastricht: MERIT.

Dankbaar, B. and Bouwman, B. (1998) *Training and Knowledge Transfer*, Brussels: CLEPA.

European Commission (EC) (1994) 'Communication on the European Union Automobile Industry', COM (94) 49 final (23 February).

Jansen, J. and Peters, J.J.W. (2001) 'Samenwerking in de Autobranche', MA thesis, University of Nijmegen.

Sandberg, A. (1995) *Enriching Production: Perspectives on Volvo's Uddevalla Plant as an Alternative to Lean Production*, Aldershot, UK: Avebury.

Womack, J.P., Jones, D.T. and Roos, D. (1990) *The Machine That Changed the World*, New York: Rawson.

13 Japanese automobile makers in Europe and the organization of the supply system

Rogier Busser and Yuri Sadoi

With the Plaza Accord and the *endaka* of 1985, Japanese foreign direct invest-ment (FDI) in South-East Asia grew quickly in the latter half of the 1980s. This new wave of Japanese investment attracted more Japanese suppliers to South-East Asia, while local enterprises also became involved in Japanese-dominated supply networks. This triggered quite some attention among a number of foremost Asian scholars, and the result was a fast-growing number of studies on the structure of Japanese supply networks in ASEAN. Trade fric-tion, a stronger yen and the quest for greater market share were the major reasons for Japanese car assemblers to establish production facilities in the United States and Europe. Although these operations have been studied from a number of different perspectives, so far little attention has been given to the formation and structure of supply systems by Japanese car manu-facturers in Europe.

The objective of this chapter is to analyse the relationship between Japan-ese car manufacturers in Europe and their Japanese and European parts sup-pliers in Europe. This analysis is made from two perspectives, namely, that of the Japanese automobile manufacturer and that of the Japanese supplier. The chapter examines whether and, if so, how Japanese automobile manu-facturers maintain their original organization of supply systems in Europe, and we look into the motives Japanese suppliers have for establishing plants in Europe. The data presented in this chapter are derived from a number of interviews and a questionnaire survey conducted by the authors.[1]

Theoretical framework

The organization of supply systems play an important role in the competitive advantage of Japanese automobile manufacturers (Dunning 1993). Numer-ous studies have been conducted on the organization of supply systems within Japan (Asanuma 1989; Smitka 1991; Asanuma and Kikutani 1992; Nishiguchi 1994; Hill 1995; Busser 2002). These and other studies emphasize the close relationship between the automaker and its suppliers in Japan, the closeness of which is evident in the fact that the asset specificity of the invest-ments of Japanese suppliers is, on average, high. On the other hand, it has been observed that Japanese automakers share more information with their suppliers and that coordination of tasks is strong, and they are therefore willing to make relation-specific investments (Dyer 1996). In many analyses of Japanese supply networks, it has been pointed out that long-term, mutually beneficial relations were one of the key factors explaining the success of the industry.

This gave rise to the question as to why Japanese enterprises were willing to cooperate in this type of long-term relationship. Some scholars opted for cultural explanations (Dore 1986). Notwithstanding the fact that many economists have difficulty in including cultural explanations in their analyses, both transaction-cost economists and game-theory adherents have done work that incorporates cultural factors in their studies to explain economically efficient organizations. This type of analysis was often used to look at Asian economies and, in particular, to explain the success of the Japanese economy. Transaction-cost economists have concluded that 'Informal constraints of Japanese society have lowered the transaction costs of adopting economically efficient organizational arrangements' and that 'the informal constraints of Japanese society do a relatively better job of holding opportunism in check than those of many Western societies' (Hill 1995). What kind of informal constraints, then, are at work here? In this chapter, we single out one informal constraint: the long-term mutually beneficial business relationship based on trust (Sako *et al.* 1998). To investigate the relevance of this informal constraint, we present two working hypotheses. First, we expect that Japanese auto manufacturers in Europe will, as far as possible, purchase parts and components from the same Japanese suppliers that supply at home. Second, we suggest that Japanese suppliers that have made investments in Europe have done so at the request of their major Japanese customers in order to sell to the Japanese car manufacturer and its group members.

Case study: the Japanese carmaker–supplier relationship in Europe

In this section, we investigate whether and, if so, how Japanese automobile manufacturers in Europe maintain their original organization of supply systems at European locations. To do so, we surveyed two Japanese automakers and sixteen Japanese suppliers located in Europe.

It goes without saying that the history of the European and Japanese automobile industries is rather different. Until rather recently, European automobile producers concentrated their efforts on having a regional presence, whereas the major global makers, such as GM, Ford and Toyota, have a global presence in the car industry. This has led to the European companies being trapped in a high cost base coupled with the fact that they have been protected from the full rigours of competition by operating in a protectionist region (Payne *et al.* 1996).

More recently, the globalization of the car industry has affected European carmakers in several ways. Of importance to this chapter is the fact that the supply chain in the European automobile industry has changed its structure drastically. Not only Just-in-Time (JIT) delivery systems but also lean production systems were implemented rather quickly in European suppliers' enterprises in the 1990s. Two more recent features of the European car industry concern the rationalization and consolidation of the supplier base. European automobile manufacturers are increasingly moving towards built-to-order systems, with suppliers continuously restructuring their operations to meet lead time between order and delivery.

The move to modular cars with fewer pre-assembled parts will rapidly change operational relationships in the supply industry. The involvement of suppliers from initial product development through to final assembly reduces product development time, cuts manufacturing expenses and improves product quality. But it also makes higher technological demands on the supplying firms. Therefore, fewer suppliers than before can survive, and the survivors will be those with higher technological capabilities. For organizational reasons, the large carmakers are reducing the number of suppliers that they purchase from, while at the same time they are fundamentally changing the way they do business with those who stay in business. Reducing the number of suppliers is a prerequisite for an improved and collaborative supplier–automaker relationship.

The total market value of the global automobile supply chain for parts and components is over US$900 billion. The number of suppliers, however, has continuously declined, from 30,000 in 1988 to less than 8,000 in 1999. In Europe, the number of first-tier suppliers decreased from 900 in 1992 to 400 in 1997 (Netherlands Foreign Investment Agency 1999).

How, given these circumstances, did Japanese automakers and their Japanese suppliers adapt to the changes that were taking place in Europe? Through our survey and interviews of two Japanese automakers and sixteen Japanese suppliers in Europe during 2000 and 2001, we found the following three major changes in their relationships in Europe.

High local content reached by Japanese automakers

Japanese automakers have achieved over 80 per cent European local content for Japanese-brand cars manufactured in Europe. It goes without saying that this figure includes products purchased from Japanese suppliers in Europe. The reasons for this rather high level of local content can be ascribed to national automobile policies, such as those of the United Kingdom and Spain. The United Kingdom implemented policies that require a minimum of 60 per cent local content for manufacturers in that country. Spain implemented a similar policy that requires 50 per cent of parts to be produced locally (Nikkan Jidosha Shinbunsha 1994).

To meet these requirements, Japanese makers started to produce engines locally, although diesel engines are mostly purchased from other European automobile makers.[2] In 1989, Honda started to produce gasoline engines in its plants in England, and in 1992 Toyota too started to produce gasoline engines in the United Kingdom. Nissan is the third Japanese car manufacturer that produces gasoline engines in the United Kingdom. Moreover, Nissan produces diesel engines in Spain (company reports), and Mitsubishi imports short engines[3] from Japan and sub-assembles them in the Netherlands.

The engine represents a high percentage of the total value of a car and also accounts for a high percentage of the total number of parts used in a car. Thus by transferring engine production to the local level, Nissan reached its 80 per cent local content rate. Nissan was being supplied by 208 local suppliers in 1998 (Fourin 2000). Toyota UK attained an 80 per cent local content rate in 1994, but in France it still hovers at around 60 per cent. Honda has been supplied by 248 suppliers, allowing an over 90 per cent local content rate (ibid.). Mitsubishi has been supplied by 174 suppliers, which has resulted in an about 85 per cent local content rate.[4]

Toyota UK spends £430 million each year on procurement, half of which is spent in the United Kingdom. Of the £215 million spent in the United Kingdom, £120 million goes to firms within a 50-mile (80-km) radius of the Burnaston plant. Toyota uses 160 component suppliers from more than ten countries (McDermott 1996). In the case of Toyota France, which started production at the beginning of 2001, about 40 per cent of parts are locally purchased. The number and role of Japanese suppliers is limited.[5]

Unlike in South-East Asia, in Europe the majority of the suppliers of the Japanese carmakers are non-Japanese. The share of the Japanese suppliers located in Europe is low. In the case of Mitsubishi in the Netherlands, among the total 174 suppliers, only fourteen are subsidiaries of Japanese suppliers in Europe. As a consequence both of a mature European automobile industry and a borderless Europe, Mitsubishi can source its components from parts suppliers located in Germany (62), France (27), the Netherlands (19) and Sweden (16).[6] In the case of Toyota too, the number of Japanese suppliers is low[7] but, unfortunately, accurate figures are not available.

Thus the role of Japanese suppliers in the local supply networks of Japanese carmakers in Europe is limited. On the other hand, the number of Japanese parts suppliers that have invested in production facilities in Europe has been increasing. As shown in Figure 13.1, as of 1998 a total of 168 Japanese suppliers had started production or were actively preparing to do so, and a number of enterprises had set up sales offices and R&D bases in Europe. Of these, fifty-six are located in England, where three Japanese automobile makers have production plants. During the peak years 1986 and 1990, fifty-four firms established their European bases. Figure 13.2 shows the number of Japanese suppliers established in Europe, categorized by host country of the investment.

Japanese suppliers in Europe: looking for a share of the European market

Why did Japanese suppliers start production in Europe? In many cases, they hoped to become more globally competitive, to gain a share of European markets, to upgrade their technologies in order to meet European standards and to achieve a global supply network. One electrical parts supplier first started a market survey office and then went on to launch production plants to establish its global parts supply network. A brake system supplier came to Europe to establish its R&D base in order to meet the high European technological and safety requirements. Our survey revealed that the major

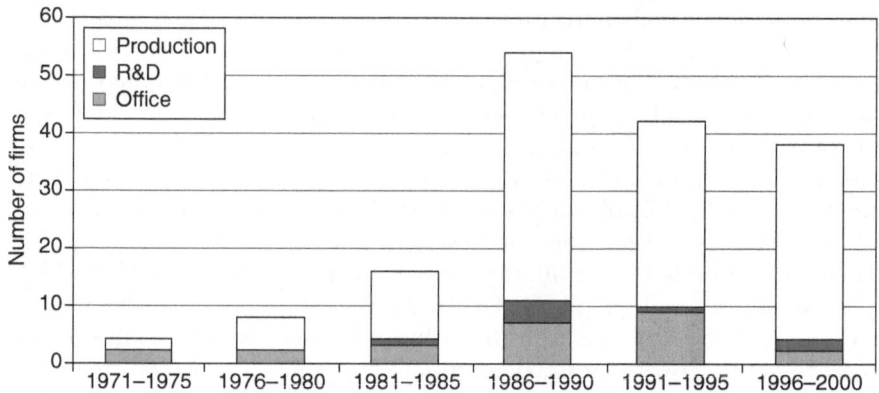

Figure 13.1 Number of Japanese suppliers established in Europe. Each column shows the number of start-ups in that year

Source: Fourin (1999)

Note
Data for 1996–2000 include established plans

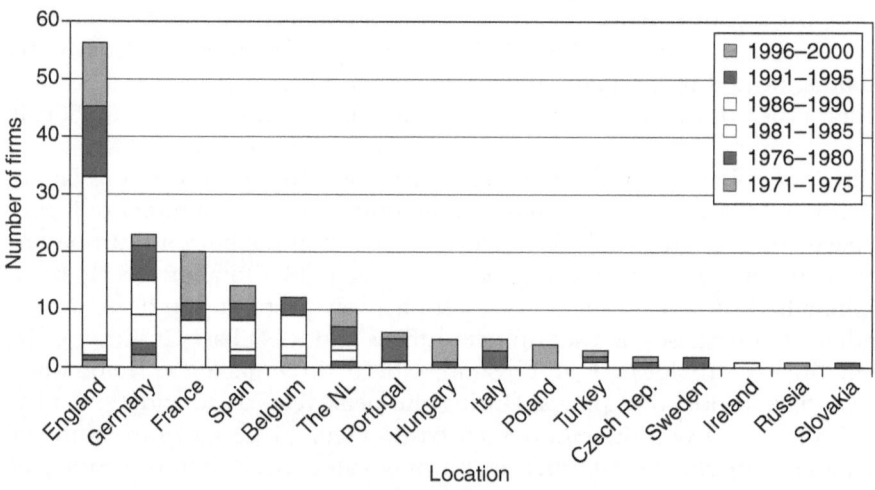

Figure 13.2 Number of Japanese suppliers established in Europe, by countries

Source: Fourin (1999)

Note
Data for 1996–2000 include established plans

Japanese customers do not request that their Japanese suppliers in Europe actually begin production in Europe.

This is very different from those cases of Japanese transplants in South-East Asian countries, where Japanese automakers have often asked, invited, or even pressed their Japanese suppliers to make investments and establish production facilities near the Japanese automobile plants. In the case of the United States, the structure and organization of the American automobile industry was such that Japanese suppliers were highly necessary. American automakers were producing nearly 90 per cent of their parts in-house. Thus there were not enough parts suppliers to supply the Japanese automakers when they started production in the United States in the early 1980s. Therefore, many Japanese suppliers were invited by their major customers to start production in the United States.

The structure of the European automobile industry was very different from the American one at the time when Japanese car manufacturers started their production facilities and investments in Europe in the late 1980s. The structure of the European automobile industry resembled the Japanese industrial structure in the sense that European automakers were outsourcing about 60 per cent of their parts. Thus high technology and high-quality parts were available from local suppliers. There was little need for Japanese automakers to request their Japanese suppliers to invest in production bases in Europe, as had previously been the case in South-East Asia and the United States.

Japanese parts producers' major purpose in starting production in Europe was twofold: to expand into the European market, and to upgrade their technology in order to meet European standards. In recent years, technological requirements for suppliers have been increasing and upgraded. Suppliers are required to be involved from the initial product development stage to final assembly. First-tier suppliers can be system suppliers that supply the bigger components related to their parts. In that case, a system supplier is in charge not only of product development, procurement of parts from the second tiers and production, but is also in charge of the quality control, and bears the product responsibility, for the whole system. This offers opportunities for the supplier to expand its business but, at the same time, it is difficult and risky to engage in the different technology areas of production.

As global parts purchasing became a common practice for automakers, the Japanese suppliers had to find ways to expand their businesses globally. Many suppliers sought new customers as automakers were (re)building global networks through mergers and alliances, such as in the cases of Renault and Nissan or DaimlerChrysler and Mitsubishi. The suppliers of Mitsubishi, for instance, were presented with a good chance to supply Daimler-Chrysler by using the network of DaimlerChrysler and Mitsubishi. Or, in the event that they were not competitive enough, they might have been replaced by the suppliers of DaimlerChrysler, even if they had been supplying parts to Mitsubishi for many years in their role as a member of a suppliers' association.

Until recently, all Japanese automakers had their own suppliers' organizations.[8] These were established in the pre-Second World War period, resurrected in the 1950s and rendered more sophisticated in the 1960s. This aspect of Japanese industrial organization increasingly gained importance from the 1970s to the 1990s. Only in the late 1990s had it become apparent that not all suppliers' organizations functioned well. Nissan restructured its suppliers' association and Mitsubishi even went so far as to terminate the suppliers' organization officially in 2002. The Japanese automaker–supplier cooperative relationship has been built up through several mutually benefi- cial activities, such as technology improvement, quality control, cost reduc- tion and so on. These activities allow suppliers to secure their long-term supply relationship with the automaker.

Despite long-term relationships, most of the suppliers are independent. For example, Denso and Aisin were once part of Toyota, but they now produce and supply more than half of their parts for other carmakers. Part- makers decide themselves whether or not to go abroad to set up a supply base for their *keiretsu* carmaker. Of course, on the other hand, it goes without saying that for a carmaker, having its suppliers close by is advantageous.

While many of the Japanese suppliers in Europe develop new customers in the European market, we observed few cases in which the Japanese car- makers and suppliers have maintained the same strong relationship as they do in Japan. Such relationships were found only in the case of products where Japanese suppliers have technological advantages, such as car air- conditioners and automatic transmissions. Cars equipped with these parts have traditionally had a low share of the European market, but have recently been gaining more market share. This relatively low demand for car air- conditioners and automatic transmissions explains why few European suppliers have invested in the production of these products, and because of this, Japanese parts suppliers were asked by Japanese automakers to invest in the European market. With the exceptions of these cases, in which Japanese suppliers and customers cooperate in a similar way in Europe as they do in Japan, Japanese suppliers are changing towards a more European style of business relations.

Differences in supplier selection between Japan and Europe

The procedure followed by European carmakers in Europe to select suppli- ers for the production of a certain part may be divided into two phases. Figure 13.3 shows two phases: (1) supplier selection, and (2) development for mass production.[9]

Phase 1: selection of parts supplier

Very much unlike what happens in Japan, carmakers in Europe usually select suppliers for each new project. Thus while long-term relations and trust rule the selection process of suppliers in Japan, this does not occur to such an extent in Europe.

As the first step, a European carmaker prepares both a 'parts drawing' (mostly a preliminary drawing only for the quotation) and a 'quotation request sheet' for each part of the new car model. The parts drawing includes the technical specifications, and the quotation request sheet shows a number of conditions, such as production volume, payment procedure and so on. The parts drawing and the quotation form are sent to a number of possible suppliers, and the purchasing department of the carmaker normally selects these suppliers.

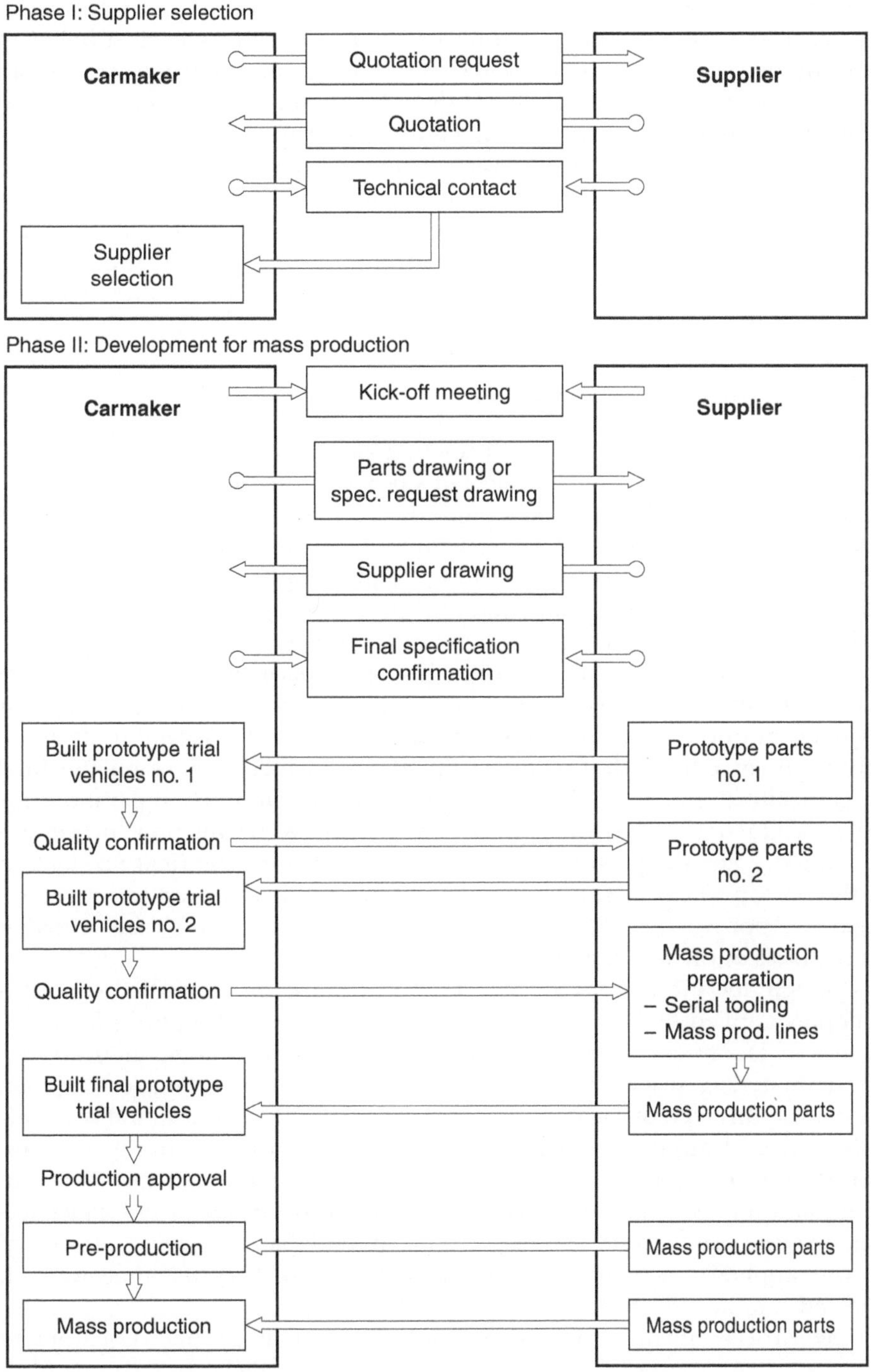

Phase I: Supplier selection

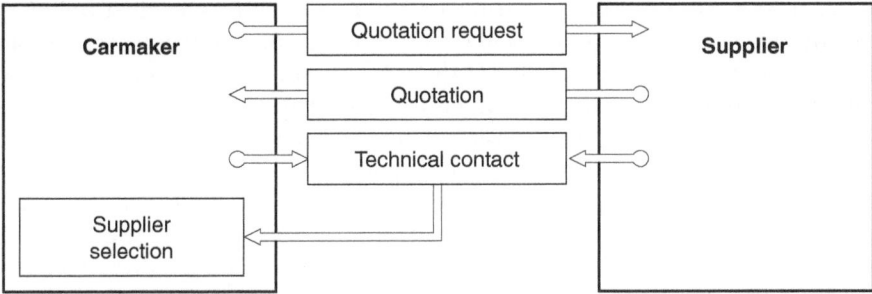

Phase II: Development for mass production

Figure 13.3 Supplier selection to mass production

Next, each supplier estimates 'unit price', 'tooling cost', 'development cost' and the 'lead time'.[10] The quotation sheet is filled in and returned to the carmaker, whereupon the carmaker has meetings with each supplier to confirm the results of the quotation sheet. In the following meetings, the carmaker expects the suppliers, with all their highly specialized technical knowledge, to present ideas on cost reduction and on quality improvement. If a particular supplier is new to the carmaker, the latter will visit the supplier's plant to check the production capabilities and quality control. Finally, after studying all conditions, particularly total cost and reliability for parts quality and delivery, the carmaker selects the best supplier. Japanese first-tier suppliers in Europe use a very similar method of supplier selection when deciding on second-tier suppliers.

Phase 2: development for mass production

After the selection procedure is finished, a kick-off meeting is held between the carmaker and each of the selected suppliers to confirm technical specifications and a development schedule for the parts. The carmaker officially releases a parts drawing. With regard to functional parts, the carmaker releases a 'specification control drawing' to the supplier and approves a 'supplier drawing' made by the supplier.

In the next step, the carmaker builds a prototype car with prototype parts obtained from the suppliers. The quality of all parts is tested at this stage and, in a minority of cases and depending on the test results, changes or adaptations in the product design have to be made. After several trials of the prototype car constructed in the plant, the carmaker decides on the final parts specifications, after which point the suppliers can start preparations for mass production. The parts made by the mass production procedure (tooling, jigs, or the production line) are delivered for the final production of the prototype car to confirm whether they can be approved for mass production. Approval for mass production should be confirmed by the carmaker on the basis of the test results of the final prototype car.

Though deviations from this selection procedure are known to occur, it is generally applied by European carmakers. Because the conditions for European and Japanese suppliers are similar, non-European companies are not discriminated against or shut out of the process. Japanese carmakers in Japan do not use this type of selection procedure for suppliers, as it would probably only increase transaction costs; however, Japanese carmakers in Europe have introduced procedures close to those of the major European carmakers. In addition, there is a crossover influence, in that this European supplier selection process has recently been introduced in Japan. In 2002, Mitsubishi Motors applied the European process in Japan for the first time when it selected suppliers for the new model Colt.[11]

The process involved in the second phase, namely the development process of mass production by the Japanese carmakers and suppliers, is almost the same in Europe as it is in Japan. Deviations can be found in a number of prototype trials. For example, some carmakers require three or more tests of prototype vehicles, while others require fewer. Unlike the supplier selection process, the development process of mass production in Japan and Europe has proceeded in a similar way.

Conclusion

In the introduction to this chapter, we presented the following working hypotheses: 'Japanese auto manufacturers in Europe will, as far as possible, purchase parts and components from the same Japanese suppliers that supply at home', and 'Japanese suppliers that have made investments in Europe have done so at the request of their major Japanese customers in order to sell to the Japanese car manufacturer and its group members'.

The results of the interviews and survey do not support these hypotheses. It was found that Japanese auto manufacturers in Europe do, in fact, make use of Japanese suppliers when these enterprises are superior to local European suppliers. Nevertheless, contrary to the situation in the United States or in South-East Asia, Japanese car manufacturers do make a large percentage of their total purchases with non-Japanese, local suppliers. The Japanese phenomenon of suppliers' organizations, which plays an important role within Japan, does not have a significant impact on the structure and organization of the supply system in Europe. Contrary to the custom in Japan, Japanese automakers in Europe apply an open bidding system for the selection of parts suppliers in Europe. However, the role of trust and the prospects for developing long-term relationships do play important roles in the process of selecting parts suppliers.

Second, we expected that Japanese car manufacturers were putting pressure on their Japanese suppliers to invest in Europe; however, our findings indicated the opposite. Most investments by suppliers were made without any request or stimulus from the major Japanese customers to do so. The reasons Japanese suppliers gave for coming to Europe were to gain market share in Europe, to become more globally competitive and to adapt to European standards. Our research also showed that Japanese partmakers have difficulties in adjusting to European business practices. Unlike the situation in Japan, partmakers cannot rely on existing relations but have to qualify each and every time to become a supplier of a certain part for a new model or car.

Notes

1 The authors interviewed at Mitsubishi Motors Europe in 2000, Toyota Motor Manufacturing France SAS in March 2001, Toyota Motor Europe Manufacturing in April 2001 and Mitsubishi Motors R&D Europe and NedCar several times during 2000 and 2001. With regard to suppliers, we received responses from sixteen out of the sixty-eight that we contacted. We interviewed eight suppliers in 2000 and 2001 (two suppliers in France, two in the Netherlands, one in Belgium, one in Germany and two at the headquarters in Japan).

2 Because of the low demand for diesel engines for small passenger cars in Japan, only a few models with diesel engines are produced in Japan. For many Japanese carmakers, purchasing from other European makers is more cost-effective than developing new diesel engines in Europe.

3 Half-built engines, meaning that parts outside of cylinder block and cylinder head are assembled in the Netherlands.
4 Interview at NedCar.
5 Interview at Toyota France.
6 Interview at NedCar Purchasing.
7 Interview at Toyota Motor Europe Manufacturing.
8 Toyota and Nissan are the most important examples of automakers that have developed their own suppliers' organizations. Mitsubishi also developed a suppliers' organization, but this was less extensive and exclusive than the Toyota and Nissan associations. Honda has never developed its own suppliers' association; the company just works with a loosely related group of suppliers (Miwa 1996). The percentage of supplier equity held by the automaker was similar for three makers in the mid-1990s: 35.9 per cent for Toyota, 32.6 per cent for Nissan and 30.2 per cent for Honda. The percentage for Mitsubishi's suppliers (42.2 per cent) was much higher than the others (Kim and Michell 1999). In Honda's case, the lower figure at least partly reflected the fact that Honda does not have a formally organized but a loosely related suppliers' association (Miwa 1996). Toyota has traditionally maintained a very close relationship with its suppliers (Cusumano 1985), and this appears to be responsible for the relatively high percentage.
9 It shows a general procedure based on interviews at both automakers (purchasing and R&D divisions) and suppliers (sales and R&D divisions).
10 The period needed to prepare mass production.
11 Interview at Mitsubishi Motors R&D Europe (December 2002).

References and further reading

Asanuma, B. (1989) 'Manufacturer–Supplier Relationships in Japan and the Concept of Relation Specific Skill', *Journal of the Japanese and International Economics*, 3 (March): 1–30.
Asanuma, B. and Kikutani, T. (1992) 'Risk Absorption in Japanese Subcontracting: A Microeconometric Study of the Automobile Industry', *Journal of the Japanese and International Economics*, 6: 1–29.
Busser, R. (2002) 'Technology Transfer through Japanese Investment in the Automotive and Electronics Industry in Thailand', in Lindblad, T (ed.) *Asian Growth and Foreign Capital: Case Studies from Eastern Asia*, Amsterdam: Askant.
Cusumano, M.A. (1985) *The Japanese Automobile Industry: Technology and Management of Nissan and Toyota*, Cambridge, MA: Harvard University Press.
Dore, R. (1986) *Flexible Rigidities: Industrial Policy and Structural Adjustment in the Japanese Economy, 1970–1980*, Palo Alto, CA: Stanford University Press.
Dunning, J.H. (1993) *The Globalization of Business*, London: Routledge.
Dyer, J.H. (1996) 'Does Governance Matter? *Keiretsu* Alliances and Asset Specificity as Sources of Japanese Competitive Advantage', *Organizational Science*, 7 (6) (November–December): 649–666.
Fourin (1999) *1999 Oshu Jidosha Buhin Sangyo* (1999 Auto Parts Industry in Europe), Nagoya: Fourin.
—— (2000) *2000 Oshu Jidosha Sangyo* (2000 Automobile in Europe), Nagoya: Fourin.
Hill, C. (1995) 'National Institutional Structures, Transaction Cost Economizing and Competitive Advantage: The Case of Japan', *Organizational Science*, 6 (1): 119–131.
Japanese Automobile Manufacturers Association (JAMA) (1999) *1999 The Motor Industry of Japan*, Tokyo: JAMA.
Kim, J. and Michell, P. (1999) 'Relationship Marketing in Japan: The Buyer–Supplier Relationship of Four Automakers', *Journal of Business and Industrial Marketing*, 14 (2): 118–129.

McDermott, M.C. (1996) 'The Revitalization of the UK Automobile Industry', *Industrial Management and Data Systems*, 5: 6–10.

Miwa, Y. (1996) *Firms and Industrial Organization in Japan*, London: Macmillan.

Netherlands Foreign Investment Agency (1999) *Automobile Industry in the Netherlands*, The Hague: NFIA.

Nikkan Jidosha Shinbunsha (1994) *Jidousha Sangyou Handbook* (Automobile Industry Handbook), Tokyo: Nikkan Jidosha Shinbunsha.

Nishiguchi, T. (1994) *Strategic Industrial Sourcing*, New York: Oxford University Press.

Payne, A.C., Chelsom, J.V. and Reavill, L. (1996) *Management for Engineers*, New York: Wiley.

Sako, M. and Helper, S. (1998) 'Determinants of Trust in Supplier Relations: Evidence from the Automotive Industry in Japan and the United States', *Journal of Economic Behavior and Organization*, 34: 387–417.

Smitka, M.J. (1991) *Competitive Ties: Subcontracting in the Japanese Automotive Industry*, New York: Columbia University Press.

14 Flexible standardization

Innovation for labour, industrial science and labour policy

Roland Springer

Criticism of the prevailing flexibility discourse

The twentieth century was, for the most part, the century of Fordism, and not only with regard to the automobile industry. The system of belt-oriented, functionally structured mass production based on a strong division of labour established by Henry Ford at the beginning of the century had an outstanding formative influence on the organizational principles of the car manufacturer not only in the United States, but worldwide. Its success was based on the possibility it created to achieve a level of productivity marked by considerable time and cost reductions through the standardizing of products, production and work sequences.

Reductions in time and costs enabled not only a continuous drop in the price of the product in relation to its enhanced performance, but also a continuous increase in wages, and thus in the purchasing power of the 'masses'. In this way, it succeeded in democratizing the car, initially a luxury product, as it were, by making it a reasonable purchase for large sections of consumers. 'Automobilization' in combination with democratization and rising consumerism in modern society are as much the results of, as the prerequisites for, the triumphal worldwide march of Fordism.

An argument could be made that the lowering of production costs could also have been achieved by means other than those chosen by Ford – namely standardization and strict division of labour; that the triumphal march of Fordism was thus not absolutely necessary, and that the history of car manufacturing could also have taken a different course. All of these objections, however rightly made in the name of contingency, do not alter the fact of Fordism's success (Noble 1986).

Dispensing with a comprehensive academic superstructure, it succeeded on a practical level in adopting the productivity precept, penetrating industrial society and combining with a specific production system. As such, the Ford method can stake its claim to be the best, a status it earned a century ago, despite both justified and unfounded criticism, and, above all, in terms of its practical effects on the progress of productivity, economic growth, the general increase in income levels per capita and on the rise in general living standards.

Until shortly before the end of the twentieth century, there was no altern-

ative production system effective enough to make a similar claim in industrial mass production. All the alternative forms of production and work
organization, especially those less based on a division of labour, have
remained, more or less, mere episodes in the more than one hundred-year
history of car manufacturing. The reasons for this vary. One is that no one
has been able to counter Ford's results with their own best-method claims to
productivity increase.

If we agree with the arguments in support of 'lean production' (Womack
et al. 1990), it would seem that, towards the end of the twentieth century, the
Toyota Production System (TPS) had best succeeded in this. However,
despite the indisputable differences, the TPS in fact more strongly resembles
the Ford system than it deviates from it (Boyer *et al.* 1998; Freyssenet *et al.*
1998). This especially applies to the ideas of the existence of, and the possible standardization of, *best methods*, adopted by both Ford and Toyota (Fujimoto 1999; Shimizu 1999), which means that progress in productivity is to be
achieved mostly through a systematic search for, and by standardization of,
the best production and work methods.

Both Ford and Toyota practise the principles of 'management by (best)
methods', i.e. the basis of 'scientific management' and thus also of industrial
science (industrial sociology, work sociology, work psychology, operations
research, business administration, etc.). According to Frederick Winslow
Taylor, who developed the concept of 'scientific management', these principles support enterprises in their search for the best production and work
methods. It is no coincidence, therefore, that Ford's methods and the discipline of industrial science peaked simultaneously, regardless of any negative
or positive orientation to either Ford's or Taylor's concepts.

Even industrial science's justified criticism of the negative effects of
Fordism and Taylorism on working conditions, namely monotony, lack of
content, occurrence of strain, etc. and of their economic limitations cannot
dispute that fact. The idea of the best or, at least, the better method has, for
the first time, really made it possible to investigate systematically and scien-
tifically all aspects of working life. As such, industrial science is a child of
Fordism and Taylorism and, like all children, over the years it has tried to
emancipate itself from its parents.

It was predominantly because of the principle of standardization that
despite its success, Fordism found itself in difficulties, even a state of crisis,
during the past ten to twenty years, and not only within the German automobile industry. Piore and Sabel, in their much-discussed study *The Second Industrial Divide* (1984), have already defined the circumstances: the bulk goods
markets, not least of which was the automobile market, in the traditional
industrial nations were widely saturated and economic improvement would
be feasible only with a narrow orientation towards customers' requests and
with a corresponding product diversification.

During the late 1980s, 'flexible automation' emerged as a catchword.
First, it enhanced future prospects: companies were able to satisfy special
requests with flexibility, such as with special orders or small-series produc-
tions. In a trend as observable now as it was then, sellers' markets seemed to

be becoming transformed into consumer markets, where flexibility requirements held far greater meaning than those of standardization. Thus since
the 1980s, the principle and concept of flexibility have been dominating
industrial and economic developments far more than the principle and
concept of standardization.

According to current observable and theoretical trends, flexibility and creativity are thought to have a formative influence on modern industrial labour
practices, and not standards and routines. In developed industrial nations,
such as the Federal Republic of Germany, standards and routines are rem-
nants of the mass production methods of the past and in fact are no longer
important in the global struggle for trade benefits, market shares and
employment levels; therefore, they no longer have a rightful place in
developed societies.

This fact is being acknowledged across the board by all parties in Germany
today, including trade unions, employers' federations, enterprises, the
media, the academic community and political bodies. The disdain of standards and routines, however, hardly stands up to the empirical realities of the
old, as well as the new, economies.

Flexibility means suppleness. Sennet (1998) was probably one of the first
to draw critical attention to what a theoretical situation of total flexibility
would mean for the social lives and the well-being of individuals both in companies and industry-wide. This particular chapter, however, concerns the
following research questions:

- Does the principle of flexibility exclude the principle of standardization,
  or do we need a new combination of both: i.e. flexible standardization?
- What are the effects of flexible standardization on an innovative job
  design?
- What are the effects of flexible standardization on the modernization of
  industrial science?
- What are the effects of flexible standardization on the modernization of
  labour policy?

Flexibility and standardization: the two sides of modernization

In consumer markets, production and work sequences are undoubtedly
under higher pressure for flexibility than sellers' markets. In a tight market
environment, there is no enterprise that can permanently afford simply to
ignore customers' specific requests concerning functionality, quantity,
quality and delivery of a product when there are competitors prepared and
able to comply with such demands at any time. A sufficient pliability of the
production and work organization of companies is, for this reason, an urgent
prerequisite for the survival and success of a business in tight and highly
competitive consumer markets.

This pliability, or suitability, of organizations enabling them to adjust to
changing environments and to follow their requirements is an apt illustration of Charles Darwin's often misunderstood theory of 'survival of the

fittest', which requires more than a pure sense of efficiency. Expressed in the language of political economics, this is the principle of market economy. The enterprises are reacting to the signals of the markets in an extremely sensible and flexible way and, to a great extent, responding to them. They would be doomed if they did not do so sooner or later, particularly if the principle of efficiency is the only criterion to which attention is paid.

The market economy, however, is not the only prevailing axiom of capitalist enterprise. Against all liberal ideology, big industrial companies in particular have no choice but to obey the methods of planned economy (Springer 1999a). Those methods subscribe to the principles of the production economy, which are no less significant for the survival abilities of companies than those of the market economy. The production economy is indeed presently focused on the most efficient possible use of the applied factors of capital and labour, and therefore includes an economic principle on labour, as well. Left alone, the market economy, i.e. the proper and punctual satisfaction of customers' requests, would lead to a pure waste of companies' resources. Only the production and the labour economy together can guarantee that capital and labour will not be wasted but used efficiently. That the pressures of the time economy, on the other hand, are subject to the principle of competition and thus to the market will not change the fact that a company acting only in tune with the market economy would not be able to achieve a sufficient level of efficiency.

Market economy and production economy are thus the Siamese twins of capitalist production, and they will be divided only at the highest risk. No automobile producer can implement the ideas expressed in flexibility discourse. If an automobile enterprise, for example, were to continue indefinitely to manufacture different cars according to customers' requests, and therefore continually change over its machinery without concerning itself with the minimization of set-up times, it would – unless it possessed a monopoly – quickly disappear from the market. The flexibility discourse that currently prevails in Germany must accept the reality that flexible production simply cannot be accomplished; however, this fact continues to be ignored on the conceptual, organizational and ideological levels.

It is assumed that enterprises can continue to be competitive only by obeying market requirements. This one-sidedness may have been forced by the fact that their flexible reactions to market requirements had been too limited by most of the varied regulations, and not only those concerning wage agreements, up to the 1990s. To make flexibility work on the practical, everyday level of operations under these restrictions, it became necessary to reinforce the concept of flexibility on the ideological level. This practical necessity, however, must not dim awareness among experts of how a modern production line really works and how it has to be set up.

The future of production and labour is, in many respects, difficult to predict. There is, however, and not only in the automobile industry, a clear, long-term trend that will be quite stable for the foreseeable future. Companies are facing stagnating world markets and are increasingly being forced to throw a greater variety of products onto the market in ever shorter periods of time, unless, however, they switch to different products and new mass markets. However, attempts in the 1980s at diversification were reversed during the 1990s instead of being reinforced.

Today's growth strategy – to return to the core business and become the best all-round supplier in a particular area – will continue into the foreseeable future. The mergers and acquisitions of these past years are the results of this strategy.

Another result is that products themselves and the production and work sequences associated with them will inevitably increase in complexity. The ability to control this process is one of the most important and difficult challenges that companies and their employees face today and into the next few years. Thus it is reasonable to suppose that demands for flexibility will continue to increase, which explains the prevailing flexibility discourse. The ability to control the increasing product and process complexity, however, cannot be attained only through flexibility. On the contrary, flexibility is, in fact, intensifying the complexity of company structures and production sequences.

Although its importance is not completely ignored, reduction in complexity should be higher on the list of priorities in today's companies. The platform strategies in use by nearly all car manufacturers today are a prominent example of this. They not only help to increase the number of items produced, thus increasing the scale economies, they are a standard measure in making sequences more easily comprehensible and controllable.

To that end, and despite all flexibility rhetoric, enterprises are applying the traditional principle of standardization, although they are denying the fact to the greatest extent possible. In the course of its new (innovative) application, however, this principle is itself being changed. For example, standardization no longer means that each car must be the same down to the smallest detail, but that, on the contrary, despite varied differences in detail, there still exist a substantial number of common elements.

'Unity in Variety' could therefore be the motto with which the principle of flexible standardization is still partly being pursued today and must continue to be pursued with regard not only to products but to processes, as well. The principle of the classical, inflexible type of standardization, on the other hand, was aimed at eliminating all variety, if possible. This might be one of the main reasons that the principle of standardization had fallen into disrepute with all the parties involved in labour policy making today. Some are afraid that entrepreneurial planning liberties will be restricted, whereas others think that standardized work will kill any creativity, especially on the part of skilled workers.

Although there is no denying these risks, companies will still have to find ways and means of reducing and controlling the complexity of their products and sequences through flexible standardization. Otherwise, they will be running the risk of sinking into the chaos of various product variants. Recently, Lacher (2001) impressively described how flexibility problems in a bulk series assembly of gears at Volkswagen AG were made more controllable by a well-directed standardization measure on the basis of team working.

Other car manufacturers, such as Ford, Chrysler-Daimler, Audi and Opel, are using the same method more and more. The teams of workers operate not only to time standards, but also to motion standards, which they develop with the support of experts from industrial engineering. Internal benchmarks of different practices help the different teams to define the best ones. These are discussed in team meetings and then depicted on standard operation sheets to be used back in the workplace. Individual deviations are not only allowed, but perceived as a basis for continuous improvement with regard to product change and tacit rationalization. The teams are organized like a kind of chamber orchestra, the members of which are integrated in the process of composition in order to find the best melody.

Innovative job design

As these examples show, enterprises are already searching for and establishing a new balance between the increasing flexibility requirements, on the one hand, and the increasing requirements for labour efficiency, on the other. Flexible standardization is thus not just a result of the new 'shareholder value economy' and of short-term thinking and action on the part of companies (Schumann 1998), but also a necessary and inevitable result of new technical and organizational requirements, which, like all innovations in the business world, are of course subject to the economic requirements and objectives of the enterprises concerned.

Against this background, the rediscovery of labour-efficient production methods in German job design is probably due to the simple fact that German enterprises, with their above-average labour costs (especially for products with unusually high labour and staff requirements) are faced with a very practical question: should they compensate for the cost disadvantages primarily by decreasing labour costs or by increasing job performance?

Flexible standardization is necessary immediately, and not only for economic reasons. It also serves as a relief to employees increasingly threatened by flexibility or creativity stress, which is unavoidable when more and more different products and product variants with ever shorter life cycles have to be produced. Currently, and not only in the automobile industry, coping with product variations is a typical feature of the normal working day, as opposed to the long-term production of a standard product. Creativity is fun, but it can lead to stress if it becomes the norm. Routine is boring, but it can also be a relief when jobs are too complex.

Luhmann (1971) was thus quite right when he referred to relieving and indispensable functions of routines, which should be taken very seriously in view of the strains that both people and organizations must bear. Standards are formalized in the form of both collective rules and institutionalized routines, relieving people and organizations from having to be permanently creative. At the same time, standards are an important prerequisite for innovation; as introductions evolve into routines, innovations arise; as innovations become routine, further creativity and additional innovation can be given fresh leeway.

Accordingly, creativity and routine, are not opposites, as commonly assumed, but rather flip sides of the same coin – namely, of innovation. Innovative job design must take this fundamental fact into consideration in both theory and practice. As yet, however, little academic research has been undertaken to probe the connection between creativity and routine further within the framework of flexible standardization and to provide an opening for new formal principles in this direction. Industrial daily operations and human resource management, as well as academic and political discussions on job designs, are suffering from the lack of investigation into this dynamic.

This is also valid not least for discussion concerning the translation of 'self-organized group work' (Gerst *et al.* 1999). For its staunch advocates, the subject of 'flexibilization' is clearly to the fore of any discussions on labour policy. Although standards and routines have always been understood to be damaging to creativity, they are also understood as being things that possibly ought not to be completely eradicated. This picks up on ideas of job design within the tradition of work humanization from the 1970s, in which a stronger enforcement of the principle of flexibility was sought in contrast with that of standardization. Within the present context of 'flexible capitalism', it is indeed doubtful whether such a position is still taking the new conditions adequately into account. Under the umbrella of the principle of flexibility, at least, standardization appears in a different light than it does under that of the principle of standardization.

The mere existence or non-existence of a repetitive nature in a job is not an adequate standard against which to measure the innovation level of any particular work form. The prevailing conviction that standardized work is conservative, while non-standardized work is innovative, is debatable. Conversely, in principle a high level of standardization obviously lends just as little support for as evidence against the innovative character of this specific work form. (Working) methods can always be innovative only with regard to defined problems, which they are able to resolve better than others can. In other words, the innovative character of a work form is determined by its potential for resolving specific problems and tasks, and not by its external features.

Today, innovations in labour policies can no longer be formed and asserted with a pure 'pursuit of happiness-approach' (Malik 2000). The conceptual and implementational hurdles in economic innovation are different and clearly tougher than they were in the 1970s, 1980s and early 1990s. Against this background, and under the conditions of a globalized consumer market and of increased pressure on labour costs, innovative job design will not be able to survive without flexible standardization. It will not be possible to achieve it without employee participation – not because the majority of workers particularly want to be involved in the process of flexible standardization, but because this process cannot be successful without their participation.

'Participative rationalization' (Springer 1999b), as it has been practised with varied success in many enterprises for more than ten years, is not yet the answer to employees' forced demands for democratization. It is not the *struggle for working conditions*, but the *struggle for jobs*, that has set the parameters for the development of the industry for more than ten years. This does not mean that the struggle for working conditions has been or will be totally given up; it rather means that the coordinates of this struggle have been altered for a long time and that the maintenance of well-paid jobs has long since dominated concerns over job design. In fact, it appears that the humanizing of job design may have dropped as low as fourth priority – after job safety, remuneration and social security – for both management and employees. Perhaps this was always the case, but these days this fact is forming the basis for labour policy productivity agreements rarely before seen in Germany.

In this context, it is significant that the call for more democracy in a company should still emerge from the classical front lines between capital and labour and from a countervailing power model in the company, neither of which, however, exists any longer in its old form. At any rate, active employee participation in rationalization derives neither from claims for democratization nor from legitimate claims for co-determination, but from the simple fact that jobs and incomes would be less safe without such participation than with it.

The active participation of the employees in the rationalization is, of course, not voluntary, but forced by the circumstances. The 'struggle for participation' (Dörre 2001), so far as it exists, on the part of the employees should not be confused with that for 'job power' (Jenkins 1975) or of 'auto-gestion' (Gorz 1967). On the contrary, the struggle for participation is not an element of class struggle that aims at the control and fundamental rearrangement of production and labour, touching on private ownership of means of production. Nobody in the industrialized world of today would propose connecting the utopia of such a rearrangement with general concepts of participative management; however, they are most probably using some of the ideas and instruments of this utopia (Wolf 1999). This is possible only because private ownership and management of the means of production is not exposed to any fundamental threat by employees and their lobbies.

Once again, this would indicate that the normative power of facts is stronger than any ideology, which likewise applies to the practice of flexible standardization, despite all the dominance of the flexibility discourse. In recent years, most automobile companies have been pursuing flexible standardization based on integral production systems with the intention of re-establishing more order in production operations (IfaA 2000), a development also noticed by the auto parts suppliers and other industries. The entire industry and all its employees must be prepared for these developments.

Innovation in industrial science

Rightly or not, the constructivistic view of the world has successfully removed the subject of standardization of methods from the agenda of serious industrial scientific research in the past ten to twenty years. This corresponded with a development in companies wherein the traditional 'management by (best) methods' was replaced by 'management by objectives', thus granting extensive choice of methods. This has applied to both management and staff levels. Work should be done according to the worker's *own* best way, and no longer by the *one* best way. Individuals should carry out their work as they each think fit, provided that the objectives agreed upon are achieved, above all those concerning quantity and quality.

The idea of choice of methods was an an answer to 'alienation of work', long criticized in the industrial sciences, a practice that was forced by Fordism and Taylorism. Essentially, it was felt that a job becomes alien to workers when the objectives and methods are dictated to them, consequently removing their opportunity to decide about which way to do their work. This results in their being in a demotivated state and detracts from their job performance. On the other hand, if workers are allowed to fix their own working methods, their identification with work will be enhanced and their motivation will increase.

This industrial-scientific axiom is the basis for all attempts at self-organization, especially with regard to the semi-autonomous teamwork implemented in automobile companies over the past years. Self-organization breaks with 'scientific management', which hinges on standardization, because it predominantly means (individual) freedom to choose production and working methods.

Under these conditions, industrial science has more or less lost its function for companies as a method-oriented discipline, because no science is necessary if individual workers decide for themselves how to do their work. Industrial science is one way of looking at organization of labour in factories; however, as a tool for analysing organizations, industrial science is not working. Why is this so?

The new form of cooperation and the new work functions involved have hitherto been less tried and tested and are not yet sufficiently interfaced methodically. They have not yet been examined adequately by industrial science. So, it is unclear, for example, to what extent workers and specialists differ from one another in terms of their having to be trained in methods of time and motion studies. Moreover, to what extent and under what conditions will workers be prepared to contribute to the standardization of their own work sequences, which certainly always entails restrictions on individual degrees of freedom? And finally, what qualification profile and job comprehension must a worker have in order both to function as a problem solver and to continue to work in a routine way?

Questions like these are making new demands on industrial science, too. Scholars are obliged not only to give academically sound answers, but also to define their separate role in the process of flexible standardization. Much-used axioms have to be tested, just as today's companies are radically questioning many of their fundamental assumptions and rules. Nobody – not even science – can credibly demand the application of these axioms to the companies today without setting a good example themselves. Innovation does not develop as a one-way street. Quite the reverse: innovations within companies initiate innovations in science, just as these lead to innovations in the company. That is why not only the organization forms and products of operational work have to be tested, but the production forms and products of scientific work, too (Schmidt 1999).

Among other things, this also means that industrial science must scrutinize the principles of flexibility and of self-organization more critically than has been done up to now, and must devote (again) more attention to the principle of standardization.

Innovation in labour policy

Standardization demands improvements from industrial science and rationalization practices. Moreover, it affects decisions about labour policy both in the company and throughout the industry. Finding themselves with an increasing level of disorder in various fields, many companies are coming back to an increased awareness of the fact that organizations without rules (standards) are not able to function effectively. At the same time – in the legitimate interest of flexibility – they do not want and are not allowed to become constricted by rigid standard corsets. That means rules and standards also have to be flexible so that they do not unnecessarily restrict the flexibility needed.

In many company agreements today, rules and standards are not rigidly fixed down to the last detail, and workers are given creative leeway. How far this should go, it is generally hard to say, and companies need to examine where necessary flexibilities should be kept and where unnecessary deviations should be prevented. Standardization degrees, once they are fixed, do not have to be set in stone, but can definitely be flexible and changeable, e.g. if important basic conditions affecting company actions are changing. This is part of the principle of flexible standardization.

Negotiations on labour policy thus normally lag behind the structural changes in the industry. Policy negotiations revolving around labour tend, at their best, to preserve the status quo or, at worst, to contribute to the deterioration of conditions. Policy negotiations are never responsible for innovation. This problem manifests itself in the lengthiness of labour political communication and decision-making processes, which are after all unavoidably becoming more complex because of the exponential increase in the number of participants involved. Just as company production and work sequences are becoming steadily more complex, so too are labour-political communication and decision procedures both in the company and industry-wide.

Today, labour-political reforms tend to take such an enormous amount of time that they risk basic conditions changing again after a decision has been made, rendering the established ruling already partly obsolete, and requiring the whole procedure to start again from the beginning. At the time when they come into effect, new rules and standards are usually already partly outdated. A comparison between the speed of labour-political innovations and the speed of industrial changes can be likened to one between a stopping train and an express train.

When labour-political communication and decision processes, however, are not capable of conforming to new situations, employees and companies will have to live with the situation, whether they like it or not. This is only avoidable if these processes and the institutions responsible for them do conform to the new conditions and, by doing so, get updated. To regard labour policy itself as a process is necessary for this kind of updating, which has to be reorganized by means of flexible standardization.

In this sense, time and operation standards should be valid not only for production work, but also for labour-political negotiations and decisions. Therefore, a process of permanent development and conformity on company and industry-wide rules should be considered. Among other advantages to this could be an increase in transparency in the rules both in companies and throughout the industry. Rules are generally created in response to modified basic conditions, however, without the old ones being withdrawn from circulation. The result is that, today, there remain a huddle of rules inside companies and throughout the industry, which really only the clever specialists can still manage to decipher. In addition, the fact that existing rules usually get duplicated further increases the opacity and effectively conveys an impression of over-regulation to those who are affected. The widely uttered call for deregulation is in part simply a response to a density of regulations that is nonsensical and nearly incomprehensible to anybody.

The principle of flexible standardization can remedy this matter if applied to the process of regulation. However, applying it in this way requires a new understanding of the regulation institutions and of their labour-political protagonists. They will have to view their tasks more as a kind of service. Above and beyond that, jobs that do not get performed on time should result in noticeable consequences not only for the organization, but for those responsible, too.

Labour policy has often been perceived by workers as a gift or service, meaning that you only get things out of it, but it is more. It is a set of rights on the one hand, and a set of duties and obligations on the other.

Conclusion

In conclusion, flexible standardization cannot be delayed. The present discussion on the future of labour suffers from the prevailing flexibility discourse in which all the protagonists in the politics of labour are more or less taking an active part. This particular discourse ignores the economic and organizational requirements of standardization that have long been in place in a company's practical operations, yet are not sufficiently taken into consideration in its organizational concepts. Economic requirements of the market and production have to be balanced anew by means of flexible standardization. The totally flexible company cannot be controlled, which would result in rationalization chaos, and flexibility and creativity stress would increase among the employees and thus hinder innovation in the long run rather than support it.

The coordinates of the discourse about the right job design are thus shifting. It is no longer the case that the traditional job design concepts such as semi-autonomous teamwork necessarily mean progress in work design; neither do they provide any promising basis for the labour-political protagonists in the company and industry-wide to act. Innovative, visionary job design concepts are those that not only make the flexibility requirements of companies and the interests for employee self-organization the priorities, but also purposefully take the production and labour-economic requirements into account.

Industrial science has to take into account the fact that flexible standardization needs theoretical approaches and design methods other than pure flexibility or pure standardization. The dominant 'pursuit of happiness approach', which is related to a struggle for working conditions, must be overcome by an approach that concentrates on the struggle for jobs.

Flexible standards are not only indispensable for work design, but also important aids in the regulation of work. Deregulation makes much sense as long as it opposes old-fashioned, inflexible rules hindering necessary flexibility. This is where the struggle against inflexible standardization finds its unlimited legitimacy. Deregulation, however, does not make sense if it results in chaos and anarchy in work sequences and labour relations inside both companies and the industry as a whole.

Flexible standardization requires a permanent process of adapting and improving the rules. Political institutions, therefore, have to reorganize their own processes of negotiation and rule setting. Above all, they have to move more rapidly and become more flexible.

References and further reading

Adler, P.S. (1993) 'Time and Motion Regained', *Harvard Business Review* (January–February): 97–108.

Badham, R. and Jürgens, U. (1998) 'Images of Good Work and the Politics of Teamwork', *Economic and Industrial Democracy*, 19 (1): 33–58.

Boyer, R., Charron, E., Jürgens, U. and Toliday, S. (eds) (1998) *Between Imitation and Innovation: The Transfer and Hybridization of Productive Models in the International Automobile Industry*, Oxford: Oxford University Press.

Dörre, K. (2001) *Kampf um Beteiligung: Herrschaft, Partizipation und Arbeitsbeziehungen im flexiblen Kapitalismus*, Wiesbaden: Westdeutscher Verlag.

Freyssenet, M., Mair, A., Shimizu, K. and Volpato, G. (eds) (1998) *One Best Way? Trajectories and Industrial Models of the World's Automobile Producers*, Oxford: Oxford University Press.

Fujimoto, T. (1999) *Evolution of a Production System at Toyota*, Oxford: Oxford University Press.

Gerst, D., Hardwig, T., Kuhlman, M. and Schumann, M. (1999) 'Group Work in the German Automobile Industry: The Case of Mercedes-Benz', in Durand, J.-P., Stewart, P. and Castillo, J.J. (eds) *Team in the Automobile Industry: Radical Change or Passing Fashion?*, Basingstoke, UK: Macmillan.

Gorz, A. (1967) *Zur Strategie der Arbeiterbewegung im Neokapitalismus*, Frankfurt: Europäische Verlagsanstalt.

Institut für angewandte Arbeitswissenschaft (IfaA) (ed.) (2000) *Arbeitsorganisation in der Automobilindustrie*, Cologne: IfaA.

Jenkins, D. (1975) *Job Power: Demokratie im Betrieb*, Hamburg: Rowohlt.

Kern, H. and Schumann, M. (1984) *Das Ende der Arbeitsteilung?*, Munich: Beck.

Kurz, C. (1998) *Repetitivarbeit – unbewältigt: Betriebliche und gesellschaftliche Entwicklungsperspektiven eines beharrlichen Arbeitstyps*, Berlin.

Lacher, M. (2001) 'Standardisierung und Gruppenarbeit – ein Gegensatz? Zum Wandel der Aggregatemontagekonzepte in der Großserienfertigung', *Angewandte Arbeitswissenschaft*, 167: 16–29.

Luhmann, N. (1971) 'Die Programmierung von Entscheidungen und das Problem der Flexibilität', in Mayntz, R. (ed.) *Bürokratische Organisation*, Cologne and Berlin: Kiepenheuer & Witch.

Malik, F. (2000) *Führen, Leisten, Leben*, Stuttgart and Munich: Deutsche Verlags-Anstalt (DVA).

Noble, D.F. (1986) *Forces of Production: A Social History of Industrial Automation*, New York: Alfred A. Knopf.

Nordhause-Janz, J. and Pekruhl, U. (eds) (2000) *Arbeiten in neuen Strukturen? Partizipation, Kooperation, Autonomie und Gruppenarbeit in Deutschland*, Munich and Mering: Institut für Arbeit und Technik.

Ortmann, G. (1995) *Formen der Produktion. Organisation und Rekursivität*, Opladen: Westdeutscher Verlag.

Piore, M.J. and Sabel, C. (1984) *The Second Industrial Divide: Possibilities for Prosperity*, New York: Basic Books.

Schmidt, G. (1999) 'Nachfrage und Angebot im Widerspruch: Anmerkungen zur anhaltenden Problematik des Anwendungsbezugs von Soziologie', in Bosch, A., Fehr, H., Kraetsch, C. and Schmidt, G. (eds) *Sozialwissenschaftliche Forschung und Praxis*, Wiesbaden: Deutscher Universitäts Verlag.

Schumann, M. (1998) 'Frißt die Shareholder-Value-Ökonomie die Modernisierung der Arbeit?', in Hirsch-Kreinsen, H. and Wolf, H. (eds) *Arbeit, Gesellschaft, Kritik: Orientierungen wider den Zeitgeist*, Berlin: Edition Sigma.

Sennet, R. (1998) *The Corrosion of Character: The Personal Consequences of Work in the New Capitalism*, London: W.W. Norton.

Shimizu, K. (1999) *Le Toyotisme*, Paris: La Découverte.

Springer, R. (1990) 'Professionalisierung der Industriearbeit? Historische Aspekte einer aktuellen Kontroverse', *Universitas*, 5: 447–457.

—— (1999a) *Rückkehr zum Taylorismus? Arbeitspolitik in der Automobilindustrie am Scheideweg*, Frankfurt and New York: Campus.

—— (1999b) 'The End of New Production Concepts? Rationalization and Labour Policy in the German Auto Industry', *Economic and Industrial Democracy*, 20 (1): 117–146.

Wolf, H. (1999) *Arbeit und Autonomie: Ein Versuch über Widersprüche und Metamorphosen kapitalistischer Produktion*, Münster: Westfälisches Dampfboot.

Womack, J.P., Jones, D.T. and Roos, D. (1990) *The Machine That Changed the World*, New York: Rawson Associates.

15 Tight flow and the competency model in the French car industry

Jean-Pierre Durand

Firms in the French car industry have gone back into the black over the past decade and a half, achieving quite exemplary levels of profitability. There are several reasons for this, but this chapter deals with only one, namely the increase in productivity gains and, more particularly, the new sources of these gains. We shall not concern ourselves here, therefore, with the scale of this increase in productivity gains, nor with the methods of calculating it, still less with the product strategies of the various car firms. I shall base my argument on the fact that considerable gains of between 5 and 12 per cent per year have been realized in the production processes under study. In general, the firms in question are large firms or networks of firms, but they include some small and medium-sized firms.

Leaving aside the technical origin of these productivity gains, I shall show how the successive reorganizations of production and work interrelate and come together in a particularly effective productive matrix. The formation of this matrix appears to have been significantly influenced by the 'Japanese productive model' or, more preciously, the 'Toyota Production System' (TPS): visits by Japanese firms during the 1980s, short-term transfers of French middle managers or engineers to Japanese plants and influences of American consultants – who have adopted the Japanese organization model – slowly modified the French car industry. There was no radical change in a short time, but step-by-step changes took place that workers and unions did not see clearly.

Each component of the productive matrix I analyse below – such as the integration effect, tight flow,[1] teamwork and the competency model – contributes to the improvement in productivity in different ways such that it is essential to analyse subtly on the basis of field observations and by deciphering managerial policies and discourse. The coherence between these elements also helps to increase the effectiveness of each factor. Finally, the involvement and mobilization of the workers' subjectivity are no longer necessarily rewarded, which is sufficient to show how the new system breaks with the Fordist employment relationship. Different French makers have kept their identity, and 'Japanization' takes different forms in different companies because it happens in different ways. Often, differences are greater between different plants of the same firm than between firms themselves. But my thesis is that even if they apply Japanese principles of production differently, the same paradigms are implemented, especially in work organization, so as to increase productivity.

Productivity gains and the emerging productive matrix

If one asks what are the bases and sources of the new productivity gains achieved over a little more than a decade, the first answer that comes to mind is of a technological nature: in both workshops and offices, the new information and communication technologies (ICTs) have made it possible to increase not only the productivity of labour, but also the global productivity of the factors of production, particularly through the rapid reduction in the cost of calculating capacity thanks to microelectronics.

However, up to now, it is not the ICTs in themselves that account for the bulk of productivity gains. Rather, these come from the reorganization of production and the reorganization of work that is linked with the technologies – that is to say, a new system for mobilizing the labour force. These reorganizations can be grouped under the concept of a new productive matrix that has arisen during the past few decades (see Figure 15.1). This concept seeks to group together into one coherent system, under tension, three elements, different in nature but functioning together in the firm – which itself has changed profoundly in structure. These elements can be

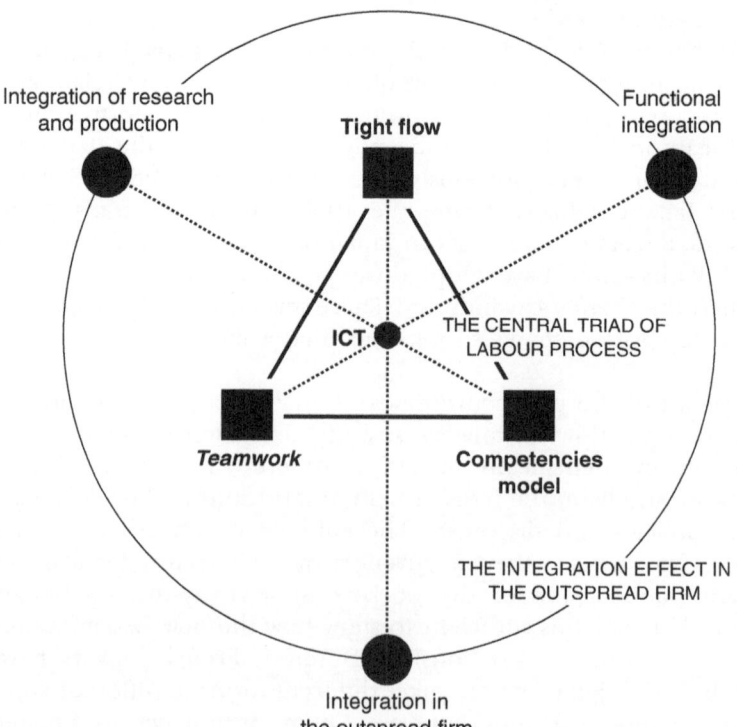

Figure 15.1 The productive matrix

represented with the help of a diagram whose three poles are the generalization of tight flow, with its consequences for the organization of work, and the introduction of teamwork and the competency model as the new system for mobilizing labour. The ICTs cut across these three poles. The reorganizations and innovations could not have taken place without them. Thus the tight flow system could not have been taken to the stage it has reached today with a management system based solely on pen and paper, since tight flow often demands instantaneous processing and communication of information. These three poles, or this central triad of the labour process (tight flow, teamwork, competence), are to be placed within the context of the reorganization and restructuring of the firm (as a network), which is closely linked with them. These transformations rest on two essential principles, the integration principle and the network principle, and are geared to integration of research and production, functional integration and integration within the extended firm (relations between the firm placing the order and the subcontractor). The changes that characterize the central triad of the labour process (tight flow, teamwork and the competency model) also have an impact on the three types of integration I have identified. In other words, one must combine the effects of the reorganization both of the centre and of its context. A careful examination of Figure 15.1 shows the close links established between functional integration and teamwork through the ICTs; the same goes for tight flow and integration within the extended firm or between integration of research and production and the competency model (the appearance of knowledge management bears witness to this close relationship).

Finally, to avoid lapsing into a sociology of work concerned only with the immediate labour process, this productive matrix must be understood as a sub-section of the post-Fordist model of capital accumulation, which cannot be presented here in all its complexity.[2] One should simply remember that the concept of production model constitutes a useful tool for organizing the heuristic to-ing and fro-ing (integrating both theoretical elements and data gained on the spot) between the micro and macro levels – without creating a pseudo intermediate level such as the firm.

The integration effect in the firm organized as a network

The term 'reticular integration' encompasses the major managerial reforms carried through over the past three decades, telescoping as it does the two concepts 'integration' and 'network'. For its strategic utility to be fully grasped, reticular integration must be broken down into three main components:

- The *integration of research and production* aims, in each firm, to bring to the market, as quickly as possible, a product or a service resulting from scientific (or other) discoveries, for a firm generally pays a lower 'admission fee' to a market if it gets into it as early as possible. It also gains an economic advantage if it creates a new niche in a market, whence comes the necessity for firms to watch for scientific, technological, strategic, or commercial developments so as to be able to integrate any such developments into the goods or services they supply.

- *Functional integration* means integration within the firm and within the plant: it means decompartmentalizing services and functions by reorganizing (functional integration) and ensuring the unhindered circulation and exploitation of computer data (data integration). The best example of this type of functional integration is *simultaneous engineering*, which enables the research department (design), the production engineers (production methods) and the workshop (manufacture) to work together and, above all, simultaneously so as to reduce the development time of a product, improve its quality and reduce its production costs. Such advantages should be the fruit of close cooperation between the various services, backed up by an intensive exchange of information drawn from common databases and used by machines and software that are already automating a considerable amount of intellectual work.
- *Integration within the firm organized as a network* is another way of dealing with the activities of the firm that are farmed out to subcontractors and suppliers, often described as 'partners'. Whereas in the past, the orders a firm placed with its suppliers were for simple products (components for a complex unit that the firm assembled itself), today the firm asks a principal supplier to design and manufacture the complex unit using components produced by secondary subcontractors. The same goes for the supply of services to the big firm. The firm, organized as a network, thus resembles a body surrounded by satellites, many of which have their own satellites around them, an increasingly complex system characterized by an intensive exchange of information. The interdependence of these actors (or integration into the 'network firm') must, in turn, be combined with the functional integration (described above) within each firm if one is to understand the nature of the organizational transformations effected over the past two decades. As we shall see, these transformations are responsible for much of the increase in time pressure on workers in work situations that have changed and in which competition (no doubt very real) is also used as a pretext for reducing the 'porosity' of working time.

Integration and the development of the network appear as forms of cooperation in work. Integration and the organization of work in network arrangements, like cooperation, are very efficient forms of work, and produce a global result much greater than the sum of its parts. However, the essential feature of the cooperation effect, as Marx showed, is that it is not paid for by the employer to his or her workers – that it is, in short, free. The same is true today for the integration effect or the network effect, which increase the productivity of each worker or, more generally, of each unit of production without the firm having to pay for the positive effect of these new forms of work.

Though all these aspects of the reorganization of the firm contribute to the increase in productivity, including in conjunction with the central triad of the labour process, it is nonetheless the latter that lies at the heart of these productivity gains and accounts for the depth of the transformations carried through over the past two decades, even if some of these changes have passed almost unnoticed.

The tight flow revolution

Ever since Henry Ford, we have known about the virtues of continuous-flow production (Figure 15.2). The speed of the conveyor belt determines the rhythm of work of the people (if the size of the labour force remains constant). People's work rate does not depend on a norm (for example, the number of items to be produced per day) imposed on them by other people. It is a machine, or a system of machines, that dictates the speed of people's work activity.

Tight flow is much more demanding than the Ford-style production flow because, over and above the work rate of the workers tied to the production line, it also mobilizes the workers who are less directly involved (supervisors and maintenance workers) who are responsible for the continuity of the flow. The tightness of the flow requires the attention and the mobilization of all the workers. Without wishing to list all the applications of tight flow organization, one can say that it is so widely used today that it has become the dominant mode of organization of production and of work, well beyond industry as such.

In the Fordist mass production system, apart from the conveyor belt, which is limited to the assembly process, each workstation produces by drawing from a stock upstream in the process and then establishing a stock downstream. In general, the organizers operate in bursts and campaigns, which means that the same thing is produced for several hours or several days, since changes in production are costly. On the basis that everything that is produced will be sold, a commercial stock supplies the final market. This system turns planning into a decisive activity that organizes the production flow, pushing the process from behind.

The tight flow system arose when production began to be regulated by demand, the idea being that the firm should manufacture only what the market had already ordered. This concept, generalized to the production

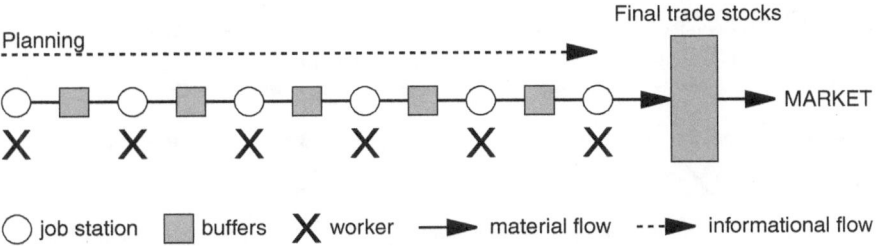

Figure 15.2 Ford-style production flow

line as a whole, means that each workstation, faced with uncertainty as to what it may be asked to supply, no longer establishes stocks as in the Fordist flow. Instead, it must be able to send the necessary products or services downstream, at the right moment (Just in Time) and in the required quantities. From this comes a double flow: a flow of material heading downstream (with no or virtually no commercial stock) and a flow of information heading upstream. To complete the picture, one must point out that there is also an information flow going downstream relating to the planning of raw material supplies and the availability of the necessary means to make production possible. But it is certainly the information flow heading upstream that pulls the process and determines what is produced (Figure 15.3).

From the economic point of view, the advantages accruing from the absence of inventory and buffers have often been highlighted:

- Capital is no longer tied up in buffers.
- There is the ability to react fully and immediately to changes in demand.
- Immediate identification of substandard or unsatisfactory products is possible, leading to a rapid solution to the problem (formerly, the entire stock might have had to be thrown away).
- Bottlenecks and malfunctions are exposed; formerly they would have been hidden by the existence of the inventory and buffers.

It seems to me, however, that though these advantages are very important, they are much less important than the organizational and political advantages that the tight flow system has brought. For the need to maintain the tight flow has major consequences from the point of view of the organization of work, cultural changes and pressures on the workers. Maintaining the tightness of the flow implies all of the following:

- Eliminating breakdowns – hence, establishing total productive maintenance (TPM). This involves a cultural change for the maintenance workers, who are no longer judged on their skill and speed at repairing a piece of machinery (something tangible, easy to measure and sometimes spectacular), but rather on their ability to prevent or anticipate breakdowns (something much less tangible from the repair operative's point of view). Paving the way for this change of mentality, the Japanese method

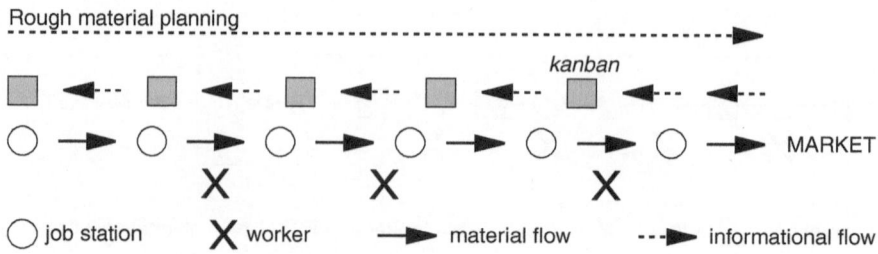

Figure 15.3 Tight flow, with no production stocks held

of the 5Ss (tidiness, order, cleanliness, getting rid of anything useless, rigour) is sometimes experienced as a sort of domestication of the workers who are being required to transform their behaviour radically.

- Producing only quality goods.
- Rapidly changing production 'campaigns' by means of the Single Minute Exchange Die.
- Permanently improving the system of production, from either the technical or the organizational point of view. This *kaizen* aims, on the one hand, at ending production hiccups and mechanical breakdowns, but, above all, its target is a permanent reduction of costs, particularly labour costs. Competition – first of all, national competition, in the case of Toyota, which created the *kaizen*, then globalization – leads management quite 'naturally' to require permanent productivity gains that can go as high as 20 per cent per year in some firms.

Over and above their purely technical functions (preventing any break in the tight flow), these socio-technical tools play a fundamental role in securing acceptance of the new productive matrix. They are, in fact, the means by which the subordinate workers are persuaded to share the firm's objectives. Who can be opposed to quality? Who can object to being able to react immediately to the market? Who can be hostile to reductions in costs (and therefore to increases in productivity) to ensure the survival of the firm? The economic objectives are transmitted through technical points during meetings of the quality circles and progress groups. General objectives, like the demands of the shareholders, are translated into detailed micro-objectives relevant to the workers, who are themselves enclosed within a restricted social and technical space. Announcing objectives that have no relation to daily life at work has little effect. But the introduction of specific practices directly linked to a concrete situation makes them tangible. In this case, TPM, *kaizen*, and total quality are the practices that correspond to globalization and the resultant competitive pressures, as well as to the demands of the shareholders of the firms in question.

In addition, these practices, which transform the relations between the shop floor workers and the technical staff (technicians or engineers), and between the former and the supervisors, put life into the participatory management system and supply it with real issues to discuss. The close relations that develop between all these social groups enable the firm to bypass the staff representatives or the union branch, or even make them redundant.

At the same time, these micro-objectives, having taken root at the local level, can give the workers new scope for autonomy, or at least give them this impression. These changes are more superficial than fundamental. The socio-productive techniques modify the discourse on work rather than the content of work (Durand *et al.* 1999). Though there are changes in the work of supervising the processes, there are far fewer changes in the direct work itself. Rather than the content of work, it is often the perception of work that has changed through the way it is presented (though one must not gloss over these presentational changes, or underestimate their social impact), for routine, repetitive work still exists, or is even expanding, in both industry and services. There are a considerable number of procedures to be followed, and the number may well increase with the development of ISO 9000 procedures. One can also show that work directives relate less and less to the tasks themselves and are pushed back further upstream, dealing rather with the objectives and the ways to achieve them (Durand 2000). Nonetheless, these are still directives, and the workers' autonomy in carrying them out remains tightly constrained by managerial imperatives.

Whether these socio-productive techniques increase the workers' autonomy a little or simply transform the way work is seen, they do tend to get workers to accept more readily the new production conditions, based on tight flow and generally characterized by increasing work rates owing to the reduced 'porosity' of work time. Putting tight flow at the centre of the analysis is therefore justified by its extension throughout industry. The reorganization of workshops into islands of production puts an end to the rather anarchic process by which parts travelled around, sometimes spending months waiting to be machined. Instead of having homogeneous machine tool shops, with the parts travelling slowly between them, components of similar types are now grouped together and machined on lines, or islands of production using machine tools dedicated to one or more types of component. The parts move from one machine to another without there being intermediate stocks (in a tight flow), following a strict schedule that the machinists have to follow. Similarly, simultaneous engineering, which combines design and production functions, means that the R&D engineers have to supply a continuous stream of information downstream to the production engineers to enable them to develop the future means of production. If one of the actors falls behind, he or she immediately causes a problem for the others. Each actor therefore has no option but to produce at the same rate as the others.

Consequently, the customer–supplier relationship appears, without this ever having been made explicit, as the practical conceptualization of the tight flow system. Not only does it apply to services (where it originated) and to industry, but it brings together all the characteristics of tight flow: Just-in-Time delivery of exactly the right quantity of goods or services of the required quality and always at the lowest possible price.

If we now turn to an analysis of the effects of tight flow on work, its first effect is to mobilize all the employees. In the tight flow organization, one can talk of a 'naturalization of constraints' in the sense that the requirements of the productive system (instructions, procedures, time pressures, emergencies, all accommodated with minimum staffing levels) are no longer mediated by a person, the boss, but are built into the need to maintain the tightness of the flow of production. Thus the constraints seem to lie outside the socio-economic world. They seem to be part and parcel of a neutral flow of materials. And like all matter, they appear to the worker as something external to him or her. The fact that the constraints are seen as something external gives them 'natural' properties, independent of human will. They are thus transformed into constraints that must be obeyed absolutely. In fact, either the worker accepts the tight flow and its associated demands, or else he or she leaves the firm. The internalization of this acceptance of what seems part of the natural order of things is very different from the social relation that the worker used to have with supervisors. Today, 'the gaffer is the flow'. Even the managers and supervisors are enslaved by the flow. Rather than commanding operations, they are caught up in administrative tasks whose main aim is to keep the flow tight. To help them in this, there is a new organization of work that fits coherently into the tight flow system of production.

Teamwork

Maintaining the tightness of the flow is a collective task. The organization of work is therefore collective, unlike in the Fordist situation where each worker was on his or her own, tied to a fixed position or even to a machine. It might be desirable to invent a new term to describe tight flow teamwork. We cannot use 'group work', because this term refers to the organization of workers in semi-autonomous groups, which originated in Scandinavia, while 'teamwork' has strong Fordist connotations. But the present system, organized around the concept of the team, is different from these earlier types of organization in the following ways:

- The end of individual workstations means that to organize the flow of production of goods or services it has to be broken down into productive units, each of which has its own work group or team.
- Each team is collectively responsible for the quality and quantity of goods produced, quantity being generally calculated not in volume but by the proportion of capacity utilization.
- The workers have had to become more and more versatile. They can and they must work in any part of the production unit for which their team is responsible. The best operatives can also be deployed in other units if the need arises. This versatility means that knowledge, and particularly the know-how that each worker has acquired in the course of his or her personal experience, is made public and socialized. This 'making public' is, moreover, perfectly consistent with the socio-technical tools presented above (TPM, total quality, *kaizen*), since these are the medium and the instruments of this process. The versatility of the workers also means that they are interchangeable. The fact that each person's know-how is made available to all and thus appropriated by the firm's management by means of increasingly more sophisticated procedures weakens the individual position of the worker. The very high rate of internal mobility (and external mobility for that matter) illustrates the increased 'replaceability' of the workers in general.
- Collective organization and collective responsibility in the team increase peer pressure on each individual. Delays, repeated absences, tiredness, or poor performances are no longer picked up or criticized by the boss but by the team, which sees its performance worsened or sees each person's share of the work increased if one of its members does not fulfil

the average norm established in the team and by the team. Cold-shouldering by the team, stress and harassment at work often originate in the teamwork system rather than being the result of an individual being hounded by a superior. It is a direct consequence of the 'natural-ization' of constraints that the weakest are forced out, since by definition the constraints cannot be evaded.

• There is now the establishment of a team leader who is one's equal, one's fellow worker gaining a little social advancement. For a slightly increased wage, and with no delegated hierarchical power, the team leaders agree to perform a contradictory role: on the one hand, control-ling their peers, and getting them to accept the management's aims as far as possible, while at the same time remaining members of their team and, as such, tied to the team's tasks and to the constraints of the flow. However, their chances of promotion are all the slimmer because the flattening of the management structure, to which they are contributing by agreeing to control their fellows without belonging to the hierarchical structure, limits the number of more senior positions available to the team leader.

To summarize, the combination of teamwork and tight flow production is fully coherent. Independent workstations could never maintain the tightness of the flow, which depends so much on the interdependence of all those who could be responsible for a breakdown. Only close cooperation between workers – collective labour – allows the tight flow to be maintained with reduced labour costs. Concern about production costs remains the priority for the management of the firm, and the question of tight flow cannot be separated from the question of reductions of costs. This, moreover, is the main lesson to be drawn from Ohno (1989), who organized Just in Time at Toyota to reduce costs (and not only to eliminate stock and buffers that were tying up capital) so that everyone would get a clear picture of all the kinds of waste generated by mass production and the inventory associated with it. The Toyota Production System (TPS), based on *kaizen*, on employees' sugges-tions, and, above all, on the permanent efforts of technical staff specialized in increasing productivity (Shimizu 1999), aims primarily at reducing labour costs. The labour thus squeezed out is then, within the Japanese system, transferred to other tasks or other factories.

To conceive of the tightness of the flow without reductions in labour costs makes no sense in a competitive (not to say capitalist) economy. The two aims go hand in hand and each reinforces the other. No doubt the flow would be still tighter if all the causes of malfunction disappeared, and if each team had sufficient human resources, quantitatively and qualitatively, to achieve its objectives. But the nature of the social system of production, capitalism, prevents this optimization, since the analysis of costs in search of maximum profits leads capitalist firms to give priority to the reduction of labour costs. What is one to do when faced with the contradiction between the tightness of the flow and reductions in labour costs? One attempts to cir-cumvent it by inventing another coherence between the organization of work in groups (teamwork) and the mobilization of the workers as agents who feel responsible for the success of the system of production. Making the workers feel responsible requires the mobilization of their subjectivity, a mobilization that is central to the competency model which has been introduced more or less throughout the world.

The competency model

In the Fordist model, the worker was paid according to the post he or she occupied, which was evaluated by technical staff according to allegedly scientific methods. The correspondence between the evaluation of the post and the worker's professional grade was an implicit admission that all the workers occupied the same post in approximately the same way. It was up to the supervisors to ensure that this was indeed the case and that 'bad elements' were removed from the post or from the firm. Pay increases were collective and were tightly linked to the policy of the state (which, in France, fixed the annual increases in the minimum wage [SMIC]) and to the collective agreements that decided on annual increases for each branch of industry. Only promotion arrangements were individualized.

During the past two decades, essentially under the combined pressure of managements of firms and employer organizations, the system of remuneration has edged towards the individualization of pay. Increasingly, it is the individual person who is remunerated and, above all, pay is determined by the way in which he or she does the job. Increasingly, managements tend to reward the degree of mobilization of each worker taken individually. This degree of mobilization is called 'behaviour' or 'competency'. Whereas a worker's qualifications, which used to be the criterion of fitness to occupy a post and therefore grading, were based on knowledge (generally acquired at school or in a training centre) and on know-how or experience (tricks and knacks picked up on the job), competency adds what we can call *knowing how to behave*. This rather ill-defined quality brings into the equation the worker's attitudes towards superiors and fellow workers. It also includes the worker's willingness to be available (overtime), responsiveness and, more generally, the worker's behaviour faced with the demands of the job.

In France, the definition of competency adopted by the MEDEF (Movement of French Firms, formerly the CNPF, the French employers' federation) is revealing in this regard:

> Professional competency is a combination of knowledge, know-how, experience, and behaviour operating in a specific context. Competency becomes apparent when put into practice in a work situation on the basis of which it can be validated. It is therefore the firm's role to identify, evaluate, and validate competency and to further develop it.

> (CNPF 1998)

In addition to the explicit inclusion of behaviour, which, as Zarifian points out (1999), has become a 'social competency', this definition asserts that only the firm can validate competency. Any indexation of wages on qualifications awarded by the national education system or by other training centres should therefore be abandoned. Competency can be evaluated and validated only in a work situation. No longer, therefore, can workers get their qualifications recognized outside the specific work situation in which they were evaluated, except where a tight labour market in a particular branch or area overturns this position. Recognition of a competency therefore has to be renewed every time a worker changes to another employer. Consequently, this definition of competency creates a major imbalance between worker and employer compared with old-style collective bargaining, since it pits the isolated worker, in search of a job and a wage, against the full power of the firm, which alone lays down the rules of the game and the criteria for evaluating competencies and behaviour.[3]

The assessment matrices that I have collected in Japan, France and the United States all provide evidence of the appearance of direct evaluation of individual behaviour. They generally consist of two main parts, of equal length. The first part deals with knowledge and know-how (qualifications) based on the evaluation of objective performance at work. The second part relates solely to attitudes and behaviour patterns. Certain rubrics can be particularly subtle in the context of *teamwork*, such as the following one used in a large Japanese motorcycle firm: 'co-operation and collaboration with others and non-attachment to one's own opinions and one's own interests'.[4] Here, the aim is to assess each individual's ability to work in a group. But other headings seem even more dubious, such as those that assess 'self-control and control of one's emotions' or 'submission to one's superior'. In France, in a firm that supplies parts for the car industry, the assessment form for workshop personnel relates to actual work performance: production, quality, safety, order and cleanliness, versatility. But the 'willingness' rubric contrasts the very good worker, who 'volunteers services before he is asked', with one who 'systematically refuses to do what he has been asked to do'. Here, it is essentially a question of working overtime in the evenings and especially on Saturdays, which a worker who has already bought a house, for example, may refuse to do. Under the 'sociability' rubric, the very good employee is seen as someone who 'doesn't make negative comments to his superior and is polite and courteous towards everyone', while the poor worker 'displays an irritable attitude to superiors, colleagues and the cleaning staff'.[5]

This assessment of behaviour is the basis of the competency model in which directives no longer relate to the tasks to be undertaken, but to the objectives to be achieved, without the necessary means (particularly in terms of staffing) always being supplied. Thus the competency model no longer checks on work, but on the loyalty of the workers, to ensure that they will put all their skill and enthusiasm into their efforts to achieve objectives that are rarely negotiable, either in terms of what the targets are or how they can be achieved. This is yet another reason why only the firm is entitled to measure competencies!

In evaluating loyalty, the competency model assesses conformity to the social norm established by the team and then demanded by the team and by the management to keep the flow tight (this reflects a certain Japanese influence). Thus assessment of the individual's aptitude for collective work and evaluation of the worker's personal commitment fit perfectly coherently with teamwork, which itself fits coherently with the tight flow. Each individual must be versatile, willing to move anywhere within the company and intellectually receptive. He or she must, in a certain sense, be dependent on the firm.

I do not share the view that this competency model necessarily leads to unhappiness at work (even though this may arise in certain circumstances[6]), but what is clear is the loss of possibilities for resistance or for trade union struggle, both of which allow space for autonomy and social games that make working conditions, even very difficult conditions, acceptable (Durand and Hatzfeld 2002). We must, however, explain what induces individuals to adopt loyal behaviour patterns in line with the management's expectations. In Japan, it is the desire to stay in the big firm, which pays its workers much better (directly in wages and, above all, indirectly in social advantages of various kinds), that accounts for this loyalty. This is what led myself and colleagues to put forward the concept of forced involvement (Durand and Durand-Sebag 1996; Boyer and Durand, 1997) to convey how little choice the workers have but to commit themselves to the objectives of the big firm if they want to stay in it. Elsewhere, particularly in Europe, insecurity of employment and fear of unemployment play the same role in leading workers to adopt the desired behaviour pattern (total loyalty) if they want to get a job or keep the one they have got. Moreover, the core–periphery model, which characterizes not only the extended firm, with its subcontractors, but also all the departments of the big firm, divides the workforce into a core, composed of very well-qualified and quite well-paid permanent staff, and a periphery made up of rather insecure workers, whose qualifications are unrecognized and who receive much lower direct and particularly indirect remuneration.

The competency model can be used to threaten the workers in the core with a possible relegation to the periphery. Only committed people are recruited into the centre, and only the most loyal are kept; they have the promise of progress or promotion in exchange for conforming to the expected behaviour pattern. The periphery, which is used, among other things, to permit an immediate adaptation to variations in the volume of demand, also has the job of inculcating loyalty in the young workers, as the condition of their promotion from insecurity to (relative) job security. The core–periphery model is a tool of social integration and/or a way of winning acceptance of the conditions imposed on the workers. For example, each person accepts an increased workload (a certain intensification of work through the reduction of the 'porosity' of working time) in exchange for the hope of improvement: advancement, or a move to a better place (escaping from the assembly line or the telephone pool in a call centre) for the workers in the core, or a permanent contract for those in the periphery.[7] As a differentiated employment structure, the core–periphery structure contributes to the introduction of the competency model and forced involvement. And these are in perfect harmony with teamwork and the tight flow.

Thus the model of the productive matrix, as defined above, is totally coherent. It leaves little or no room for opposition or resistance, since individual assessment is based essentially on loyalty. Not only does the model rule them out, but, at the same time, it develops practices conducive to social integration through participatory management and the socio-productive techniques associated with tight flow.

Conclusion: stress and optimism

This chapter demonstrates the systemic coherence of the new productive matrix based on tight flow, teamwork and the competency model. Can this coherence overcome the contradictions between workers and employers and keep French car manufacturers profitable over the long term?

On the one hand, this is the model with which the car industry in France emerged from its crisis at the beginning of the 1980s, with borrowings from the Japanese model, so buoyant earlier through the introduction of lean production as systematized by the Massachusetts Institute of Technology (MIT). It may, therefore, seem to be under threat when the conditions for the maintenance of forced involvement disappear, i.e. with a return to virtually full employment, which would considerably reduce competition between workers; corporate welfare, which encourages Japanese workers to stay in the big firm, has little influence on French workers. These could, therefore, revert to forms of resistance and, for example, *simulate* loyalty to satisfy managerial demands while inwardly adopting a totally opposite attitude. If that happens, this new productive matrix will have been nothing more than a transitional model on the way to something else, as yet unknown.

But looking at it another way, one can show that during its last crisis of accumulation and, more precisely, in order to confront this crisis, the employers discovered new methods (at least new compared with the Golden Age[8]), such as the systematization of job insecurity. Employers invented structural insecurity by generalizing the core–periphery model and extending it to the very heart of their operations (management of the labour force, management of capital, product design, etc.). From this point of view, during the crisis the employers seem to have learned much more than the workers, since, at the end of this crisis of accumulation, the balance of power between capital and labour seems far more unfavourable to labour than in the Golden Age: to mention just one example, the share-out of added value, labour's share would seem to have gone down by 6 per cent in France between 1979 and 1998. But, paradox of paradoxes, it is in periods of economic growth, when workers are under least pressure, that they learn most: the present period could therefore lead to the invention of new forms of social struggle that would once again modify the productive matrix.

Notes

1 Tight flow is more than Just in Time, but it is on the same principle: to organize a continuous flow of matter (and sometimes of information) within and between companies. JIT is only a technique to produce and deliver items of good quality and quantity at the right time. The tight flow is a *paradigm* (see page 275) which

organizes the discipline (and the path) of work through the continuous flow of all activities (including intellectual work). For instance, the customer–supplier relationship applied everywhere today is an illustration of the general implementation of the tight flow paradigm.

2 Cf., for the presentation of the developing model of production, Durand (1999).
3 I repeat, this imbalance characterizes the general situation, but not in tight segments of the labour market in which the firm is subject to wage pressure from those selling their labour power. This situation was also true of the Fordist collective bargaining arrangements (cf., for example, the first chapter, on the presentation of the big socio-economic changes in Sweden, in Durand 1994).
4 Taken from an internal document of a large Japanese motorcycle company (no date).
5 Taken from an internal company document, 'individual assessment sheet' of a French firm.
6 On my critique of the extreme thesis of ontological suffering, at work in particular, see Durand (2000a).
7 We find here the whole subtle interplay between internal and external flexibility, which cannot be understood without the problematic of the search for mobility by workers who seek to get the best out of a structure that dominates them.
8 Translator's note: the author refers here to the *Trente Glorieuses*, the name given in France to the long period of sustained growth roughly stretching from the late 1940s to the early 1970s.

References and further reading

Bollier, G. and Durand, C. (eds) (1999) *La Nouvelle Division du travail*, Paris: L'Atelier.

Boyer, R. and Durand, J.-P. (1997) *After Fordism*, London: Macmillan.

CNPF (1998) *Objectif compétences*, vol. 1, Journées internationales de Deauville.

Coutrot, T. (1998) *L'Entreprise néo-libérale, nouvelle utopie capitaliste?*, Paris: La Découverte.

Durand, J.-P. (ed.) (1994) *La Fin du modèle suédois*, Paris: Syros.

—— (1999) 'Le Nouveau Modèle productif', in Bollier, G. and Durand, C. (eds) *La Nouvelle Division du travail*, Paris: L'Atelier.

—— (2000a) 'Combien y a-t-il de souffrance au travail?', *Sociologie du Travail*, 3: 313–340.

—— (2000b) 'Les Enjeux de la logique compétences', *Gérer et Comprendre, Annales des Mines*, December: 16–24.

Durand, J.-P. and Durand-Sebag, J. (1996) *The Hidden Face of the Japanese Model*, Melbourne: Monash Asia Institute.

Durand, J.-P. and Hatzfeld, N. (2002) *Living Labour: Life on the Line at Peugeot France*, London: Palgrave.

Durand, J.-P., Stewart, P. and Castillo, J.-J. (eds) (1999) *Teamwork in the Automobile Industry: Radical Change or Passing Fashion?*, London: Macmillan.

Duval, G. (1998) *L'Entreprise efficace à l'heure de Swatch et McDonald's*, London: Syros.

Linhart, D. (1994) *La Modernisation des entreprises*, Paris: La Découverte.

Ohno, T. (1989) *L'Esprit Toyota*, Paris: Masson.

Rochefort, T. (2000) 'Invention du travail et nouvelles combinaisons productives efficaces', *Issues*, 55–56: 99–128.

Shimizu, K. (1999) *Le Toyotisme*, Paris: La Découverte.

Zarifian, P. (1999) *Objectif compétence*, Paris: Editions Liaisons.

Index